Microscales of Turbulence

Microscales of Turbulence

Heat and Mass Transfer Correlations

Vedat S. Arpacı

The University of Michigan
Ann Arbor

Gordon and Breach Science Publishers

Australia • Canada • China • France • Germany • India • Japan •
Luxembourg • Malaysia • The Netherlands • Russia • Singapore •
Switzerland • Thailand • United Kingdom

Amsteldijk 166
1st Floor
1079 LH Amsterdam
The Netherlands

British Library Cataloguing in Publication Data

Arpacı, Vedat S. (Vedat Salih), 1928–
    Microscales of turbulence : heat and mass transfer
    correlations
    1. Fluid dynamics 2. Thermodynamics 3. Energy transfer
    I. Title
    620.1'064

ISBN 90-5699-565-0

For
Mustafa Kemal
Gone, but not forgotten

Imagination is more important than knowledge

—Einstein

CONTENTS

PREFACE

This book presents the microscales of complex—buoyant, thermo-capillary, two-phase, reacting, radiating, pulsating—turbulent flows and interprets heat and mass transfer correlations in terms of these scales. The content is an outgrowth of invited lectures at several universities in the United States and keynote lectures I delivered at a number of international conferences. Manuscripts based on these lectures are in the archival literature.

Raison d'être for the work is to bridge the gap existing between statistical/computational studies of turbulence and experimental data correlations of turbulent heat and mass transfer.

I thank Dr. Rodney Tabaczynski of Ford Scientific Laboratories for my introduction to microscales and support with several grants during the past two decades.

Research leading to chapter 7 was carried out at the Combustion Research Facility of Sandia National Laboratories in Livermore, California, while I was on sabbatical leave from The University of Michigan and holding a Department of Energy fellowship through the Associated Western Universities and a research grant from the Gas Research Institute.

The work could not have been completed without the help of my student Shu-Hsin Kao who, with unusual dedication, converted my lecture notes on microscales to the present form of this book. I am also grateful to my students Chih-Yang Li and Laila Guessous for comments on the manuscript.

CHAPTER 1

INTRODUCTION

Increased attention is being paid to turbulent flows for more than a century but to date a general approach to their solutions has not been found and is not expected to be discovered in a foreseeable future. One major stumbling block to progress is our failure to solve nonlinear partial differential equations which are assumed to govern turbulent flows. Because of rapidly increasing computational capabilities, recent attention is increasingly being paid to solution of these equations by direct numerical or large eddy simulations. Despite extensive computational efforts, however, accurate quantitative predictions are still difficult to make without heavily relying on some ad hoc assumptions. In short, the analytical and/or computational tools available for turbulence are so far not able to provide the whole story. One also needs intuitive use of dimensional arguments to complete the story between mathematics and actual flows. An inspection of experimental literature clearly shows that some aspects of turbulence depend only on a few transport properties (such as diffusivities, dissipation rates, etc). In such situations dimensional methods readily lead to an answer except for a numerical coefficient. Outstanding examples are equilibrium range of energy and entropy spectra, and heat/mass transfer correlations. Here, an illustrative example, would clarify this fact further. First, note that *analytical tools for linear problems are quite different than those for nonlinear problems but the same dimensional tools equally apply to both problems.* Accordingly, we use an old linear problem (semi-infinite, stagnant, viscous fluid suddenly accelerated from boundary) for our demonstration (Fig. 1.1). Conversion of the governing equation,

1

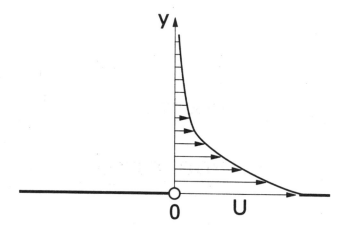

FIGURE 1.1:

$$\frac{\partial u}{\partial t} = \nu \frac{\partial^2 u}{\partial y^2} \, ,$$

(by using Fourier transforms in space or Laplace transforms in time or a similarity variable) to an ordinary differential equation eventually leads to the exact solution,

$$\frac{u}{U} = \text{erfc} \left(\frac{y}{2\sqrt{\nu t}} \right)$$

which yields, for the friction coefficient,

$$f = \frac{2}{\sqrt{\pi}} \left(\frac{\nu}{tU^2} \right)^{1/2} . \tag{1.1}$$

Now, suppose we do not have the analytical tools (which, as we already mentioned, is the case for turbulent flows) to solve the problem. Consider an approximate solution, say an integral formulation coupled with an assumed velocity profile over boundary layer δ (Fig. 1.2)

$$\frac{d}{dt} \int_0^\delta u \, dy = \nu \left(\frac{\partial u}{\partial y} \right)_w \, ,$$

and

$$\frac{u}{U} = \left(\frac{y}{\delta} \right)^2$$

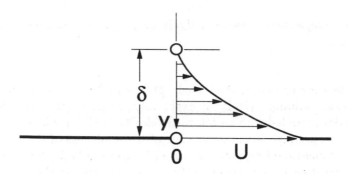

FIGURE 1.2:

which gives the boundary thickness,

$$\delta = \sqrt{12\nu t} \tag{1.2}$$

and the friction coefficient,

$$f = \frac{2}{\sqrt{3}}\left(\frac{\nu}{tU^2}\right)^{1/2}. \tag{1.3}$$

Next, suppose that we do not even have the tools to construct an approximate velocity profile (which is the case for turbulent flows). On dimensional grounds, however, from the differential formulation,

$$\frac{U}{t} \sim \frac{U}{\delta^2},$$

we have the boundary layer thickness,

$$\delta \sim \sqrt{\nu t}$$

and the friction coefficient,

$$f = \frac{2}{C}\left(\frac{\nu}{tU^2}\right)^{1/2}.$$

Thus, by dimensional arguments alone, we get the friction coefficient except for numerical coefficient C.

Clearly, the friction coefficient, obtained by three different methods, has a *universal* part,

$$\left(\frac{\nu}{tU^2}\right)^{1/2}$$

which is independent of these methods, and it has a *structural* (numerical) part,

$$\frac{1}{C}$$

which depends on the method used. Differential formulation, leading to the exact solution, yields the exact value of $C = \sqrt{\pi}$. An integral formulation, coupled with an assumed velocity profile, leads to $C = \sqrt{3}$ which is surprisingly accurate in view of the crudeness of the assumed profile. *Dimensional approach*, along with other methods, *takes care of the universal part but is silent to the numerical value of the constant.*

Because of its relevance to our objective, let us also consider a steady problem related to the foregoing example. Relative to an observer moving with velocity U, the problem becomes

$$U\frac{\partial u}{\partial x} = \nu\frac{\partial^2 u}{\partial y^2} \; ; \quad u(0,y) = U \; , \;\; u(x,0) = 0 \; , \;\; u(x,\infty) = U \; .$$

Each solution, corresponding to the three methods considered above, is readily obtained by letting

$$t = \frac{x}{U} \; .$$

In the case of dimensional approach, for example, the temporal boundary layer becomes a spatial boundary layer,

$$\delta \sim \sqrt{\frac{\nu x}{U}} \; , \tag{1.4}$$

and the friction coefficient yields

$$f = \frac{2}{C}\left(\frac{\nu}{Ux}\right)^{1/2} \; . \tag{1.5}$$

For reasons to be clear in Chapter 4, let us consider the rate of mechanical energy related to the foregoing steady problem,

$$u\left(U\frac{\partial u}{\partial x} = \nu\frac{\partial^2 u}{\partial y^2}\right) \; .$$

Continuing on dimensional grounds, letting $U \sim u$, $x \sim \ell$ and $y \sim \delta$, we get

$$\frac{u^3}{\ell} \sim \nu\frac{u^2}{\delta^2} \sim \mathcal{P} \; , \tag{1.6}$$

and

$$u \sim (\mathcal{P}\ell)^{1/3} \sim \delta(\mathcal{P}/\nu)^{1/2} \, ,$$

or

$$\delta \sim \ell^{1/3} \left(\frac{\nu^3}{\mathcal{P}} \right)^{1/6} . \tag{1.7}$$

Suppose we have a region of flow (to be clarified in Chapter 4) for which both scales approach one scale,

$$\left(\begin{matrix} \delta \\ \ell \end{matrix} \right) \to \delta_K \, ,$$

and Eq. (1.7) is reduced to

$$\delta_K \sim \left(\frac{\nu^3}{\mathcal{P}} \right)^{1/4} . \tag{1.8}$$

Then, with an inertial estimate for $\mathcal{P}$ from Eq. (1.6),

$$\frac{\delta_K}{\ell} \sim Re^{-3/4} \, , \quad Re = \frac{u\ell}{\nu} . \tag{1.9}$$

For $Re = 10^6$ and $\ell = 3$ cm (a geometric scale), Eq. (1.9) yields

$$\boxed{\delta_K \sim 1 \, \mu\text{m}} . \tag{1.10}$$

What so far we considered (by somewhat loose arguments in the preceding few pages) are amplified and justified in the rest of this monograph. *Our objective is to obtain, by a dimensional approach, the microscales of complex* (buoyancy and surface tension driven, two-phase, reacting, radiation affected, unsteady, etc) *turbulent flows and interpret so-far assumed empirical heat and mass transfer correlations in terms of these scales.*

1.1   ORIGIN OF MICROSCALES

First, eliminating time next length between the fundamental units of two kinematic concepts, the dissipation rate of the mean kinetic energy per unit mass of turbulent fluctuations,

$$[\epsilon] \equiv \text{L}^2/\text{T}^3 \, , \tag{1.11}$$

and the kinematic viscosity,

$$[\nu] \equiv L^2/T , \qquad (1.12)$$

Kolmogorov (1941) originated a length scale and a time scale

$$\eta = \left(\frac{\nu^3}{\epsilon}\right)^{1/4} , \quad \tau = \left(\frac{\nu}{\epsilon}\right)^{1/2} , \qquad (1.13)$$

and from their ratio, a velocity scale,

$$v = (\nu\epsilon)^{1/4} \qquad (1.14)$$

which are called now the microscales of isotropic turbulent flow. The intuitive idea leading to these scales was later extended by Oboukhov (1949) and Corrsin (1951) to a thermal scale for fluids with vanishing Prandtl number,

$$\eta_\theta^C = \left(\frac{\alpha^3}{\epsilon}\right)^{1/4} , \quad Pr \to 0 \qquad (1.15)$$

which results from the elimination of time between ϵ and the thermal diffusivity,

$$[\alpha] \equiv L^2/T . \qquad (1.16)$$

Kolmogorov's intuitive idea, however, no longer holds for the last of the length scales,

$$\eta_\theta^B = \left(\frac{\nu\alpha^2}{\epsilon}\right)^{1/4} , \quad Pr \to \infty \qquad (1.17)$$

proposed by Batchelor (1958) for fluids with large Prandtl numbers, which requires a combination of three, rather than two, thermo-mechanical transport properties. Clearly, a general methodology is needed not only for recovery of the foregoing scales but also for discovery of the microscales of more complex (buoyancy driven, pulsating, two-phase, reacting, radiating, etc.) flows. *First objective* of the present monograph is an introduction to this methodology by following the recent studies of Arpacı (1986, 1990a, 1995a,b). Although the foregoing three length scales have been extensively used in studies on energy and entropy spectra, their relevance to turbulent heat and mass transfer correlations have been apparently overlooked except for the Arpacı studies. *Second objective* of the monograph is to demonstrate the microscale foundation of the existing and so-called empirical heat and mass transfer correlations, develop models for future correlation of experimental data, and bridge the gap between heat/mass transfer and fluid mechanics studies. Although the dissipation of mechanical (kinetic) energy is well-understood, that of thermal energy is so-far left untreated in classical thermodynamics and is the concern of the next section.

1.2 ENERGY DISSIPATION-ENTROPY PRODUCTION

1.2.1 Thermodynamic Foundations[1]

Consider a reciprocating engine. For each cycle of this engine, the *First Law* states (Fig. 1.3a)

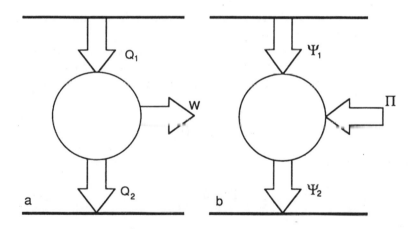

FIGURE 1.3: Two Laws of Thermodynamics. (a) First Law (b) Second Law (proposed).

$$Q_1 - Q_2 = W \,. \tag{1.18}$$

[1]One gets used to thermodynamics rather than learning it.

Planck

The usual approach to the *efficiency* of this engine operating under the *Carnot cycle* leads to a scale for the *absolute temperature* (Kelvin)

$$\frac{Q_1}{Q_2} = \frac{T_1}{T_2} \, , \tag{1.19}$$

and to the definition of *entropy flow*,

$$\Psi = \frac{Q}{T} \, . \tag{1.20}$$

The classical *Second Law* for an engine states that

$$\frac{Q_1}{T_1} \leq \frac{Q_2}{T_2} \, , \tag{1.21}$$

or,

$$\Psi_1 \leq \Psi_2 \, , \tag{1.22}$$

the equality being for the (reversible) Carnot cycle and the inequality being for an *irreversible* cycle.

This inequality suggests, from the view of a *balance principle*, an *entropy production* Π which is a measure for the *energy dissipation*. In terms of this production, the Second Law for any cycle may be expressed as (Fig. 1.3b)

$$\Psi_2 - \Psi_1 = \Pi \, . \tag{1.23}$$

For the Carnot cycle,

$$\Pi = 0 \, , \tag{1.24}$$

and

$$\Psi_1 = \Psi_2 = \text{Const.}$$

which is identical to Eq. (1.19).

1.2.2   Thermal Displacement-Deformation

To illustrate the energy dissipation–entropy production relation, consider a system (Fig. 1.4). For an infinitesimal process excluding any

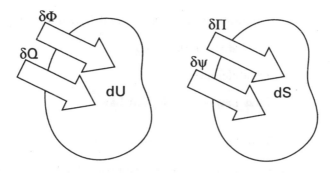

FIGURE 1.4:

work but including an imposed energy generation, the First Law applied to this system gives

$$dU = \delta Q + \delta \Phi \tag{1.25}$$

U being the internal energy, Q the heat flow and Φ the energy generation from an external source. The Second Law for the system is

$$dS = \delta \Psi + \delta \Pi \tag{1.26}$$

S being the (internal) entropy, Ψ the entropy flow and Π the entropy production.

> Note the formal similarity between Eqs. (1.25) and (1.26), from the view of the conservation of a property B,
> $$dB = \delta F + \delta G \,,$$
> F and G respectively being the flow and generation of B. Here d denotes an infinitesimal volumetric change and δ an infinitesimal change in flow across boundaries.

Now, rearrange the First Law by expressing the heat flow in terms of entropy flow,

$$dU = \delta(\Psi T) + \delta \Phi \,, \tag{1.27}$$

and multiplying by temperature, state the Second Law in energy units,

$$TdS = T\delta\Psi + T\delta\Pi \, , \tag{1.28}$$

and consider the energy difference

$$\text{First Law} - T \times \text{Second Law} \, ,$$

$$dU - TdS = \Psi\delta T + \delta\Phi - T\delta\Pi \, . \tag{1.29}$$

For a reversible process, every term on the right-hand side of Eq. (1.29) is naught, and

$$dU - TdS = 0 \tag{1.30}$$

which is the *Gibbs thermodynamic relation*. Also

$$T = \text{Constant} \tag{1.31}$$

for this process, and Eq. (1.30) in terms of the Gibbs function,

$$G = U - TS \, , \tag{1.32}$$

leads to

$$dG = 0 \, . \tag{1.33}$$

For an irreversible process *assumed* locally reversible, Eq. (1.30) continues to hold and

$$0 = \Psi\delta T + \delta\Phi - T\delta\Pi \tag{1.34}$$

which gives

$$\delta\Pi = \frac{1}{T}\left(\Psi\delta T + \delta\Phi\right) \, , \tag{1.35}$$

or,

$$\text{Entropy production} \equiv \frac{1}{T}\left(\text{Energy dissipation}\right) \tag{1.36}$$

In Eq. (1.27) we expressed the heat flow of the First Law in terms of the entropy flow,

$$\delta Q = \delta(\Psi T) = T\delta\Psi + \Psi\delta T \, . \tag{1.37}$$

Note that only a part of the heat flow, $\Psi\delta T$ of Eqs. (1.35) and (1.37), produces entropy. Assuming entropy flow to be a "thermal force" and temperature to be a "thermal length,"

$$\delta Q_D = T\delta\Psi \qquad (1.38)$$

may be interpreted as "thermal displacement" which produces no entropy, and

$$\delta Q_L = \Psi\delta T \qquad (1.39)$$

as "thermal deformation" which produces entropy by irreversible conversion (dissipation) into internal energy.[2] Thus, Eq. (1.35) becomes, in terms of Eq. (1.39),

$$\delta\Pi = \frac{1}{T}\left(\delta Q_L + \delta\Phi\right) , \qquad (1.40)$$

and Eq. (1.37), in terms of Eqs. (1.38) and (1.39),

$$\delta Q = \delta Q_D + \delta Q_L . \qquad (1.41)$$

Foregoing concepts, although very important in turbulence studies, are overlooked in classical thermodynamics which mainly deals with reversible (equilibrium) processes. That is, $T = $ Constant, $\delta T = 0$ and $\delta Q_L = 0$.

1.2.3   Second Law Alternatives

Consider a thermomechanical process through an infinitesimal control volume. The First Law applied to this control volume gives (Fig. 1.5a)

$$dH^0 = \delta Q - \delta W , \qquad (1.42)$$

where

$$H^0 = U + pV + U_K + U_P \qquad (1.43)$$

is the stagnation enthalpy, V the volume, U_K and U_P the kinetic and potential energies. After a sign change shortly to be clear, let δW be split, in a similar manner to δQ, as[3]

$$\delta W = -\left(\delta W_D + \delta W_L\right) , \qquad (1.44)$$

---

[2]Subscript L of Q_L will become clear in the next subsection.
[3]The explicit tensorial form of δW is left to Section 1.3.

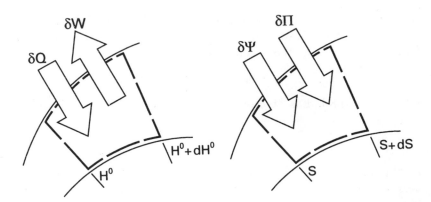

FIGURE 1.5: Two Laws for a differential control volume.

where δW_D is the displacement work which produces no entropy and δW_L is the deformation work which, by irreversible conversion (dissipation) into internal energy, produces entropy, and is called *lost work* in classical thermodynamics. Then, Eq. (1.42) becomes

$$d(U + U_K + U_P) = \delta(Q_D + Q_L) - d(pV) + \delta(W_D + W_L) \quad (1.45)$$

(the sign change introduced in Eq. 1.44 reflects the opposite sign convention for pressure and shear stress).

The mechanical energy balance, obtained either by eliminating thermal effects from Eq. (1.45) or from the mechanical energy associated with Newton's Law, is

$$d(U_K + U_P) = -V\,dp + \delta W_D \ . \quad (1.46)$$

As is well-known, Eq. (1.46) is reduced to the Bernoulli equation for steady, inviscid flow of an incompressible fluid.

Now, for a thermomechanical process, consider the energy difference (Fig. 1.6)

First Law − Mechanical Energy − T(Second Law) ,

or,

(Total − Mechanical − Thermal) × Energy

which, in terms of Eqs. (1.45), (1.46) and (1.26), yields

$$dU + p\,dV - T\,dS = \delta Q_L + \delta W_L - T\delta\Pi \ . \quad (1.47)$$

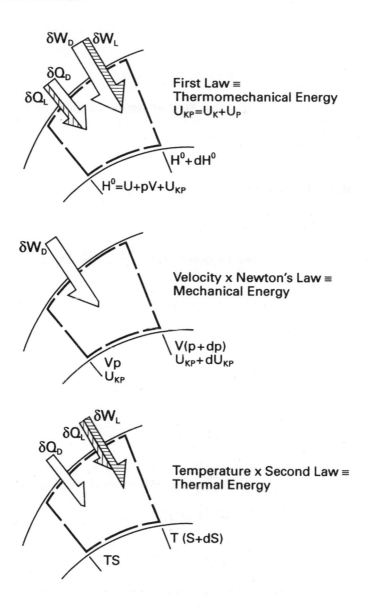

FIGURE 1.6: First Law − Velocity × Newton's Law − Temperature × Second Law.

For a reversible process, every term of the right-hand side of Eq. (1.47) is naught, and

$$dU + p\,dV - T\,dS = 0 \qquad (1.48)$$

which is the *Gibbs thermodynamic relation* (recall Eq. 1.30). Also p and T are constants for this process, and Eq. (1.48), in terms of the Gibbs function,

$$G = U + pV - TS \qquad (1.49)$$

leads to

$$dG = 0 . \qquad (1.50)$$

For an irreversible process assumed *locally* reversible, Eq. (1.48) continues to hold and

$$0 = \delta Q_L + \delta W_L - T\delta\Pi \qquad (1.51)$$

which gives

$$\delta\Pi = \frac{1}{T}\left(\delta Q_L + \delta W_L\right) , \qquad (1.52)$$

or

$$\text{Entropy production} = \frac{1}{T}(\text{Heat \& work dissipation into internal energy})$$

Now, the Second Law given by Eq. (1.26) may be given alternative forms in terms of thermal deformation and deformation work as

$$dS = \delta\Psi + \frac{1}{T}\left(\delta Q_L + \delta W_L\right) , \qquad (1.53)$$

or, in view of Eq. (1.38), as (Fig. 1.7a)

$$dS = \frac{\delta Q_D}{T} + \frac{1}{T}\left(\delta Q_L + \delta W_L\right) , \qquad (1.54)$$

or, in view of Eq. (1.41), as (Fig. 1.7b)

$$dS = \frac{\delta Q}{T} + \frac{\delta W_L}{T} . \qquad (1.55)$$

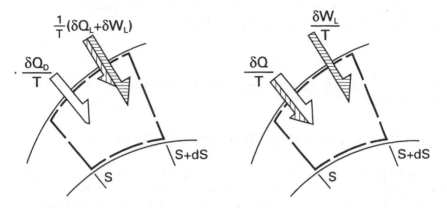

FIGURE 1.7: Two alternatives of Second Law for a differential control volume.

The explicit forms of δQ_L and δW_L will be given in the next section which deals with the rate of First and Second Laws of Thermodynamics.

When the First Law includes all (heat, work, radiative, electromagnetic, chemical, and nuclear) forms of energy, Entropy production is generalized to

$$\frac{1}{T}(\text{All forms of energy dissipation into internal energy}) ,$$

and Eqs. (1.53) and (1.55) respectively become

$$dS = \delta\Psi + \frac{1}{T}(\text{All forms of energy dissipation}) , \qquad (1.56)$$

and

$$dS = \frac{\delta Q}{T} + \frac{1}{T}(\text{All forms of energy dissipation excluding heat}) . \qquad (1.57)$$

Note the two sources of dissipation, diffusion and hysteresis, the latter being not our concern. Let the diffusive internal energy, heat, and work associated with the optical limit of electromagnetics (gas radiation) be U^R, Q^R and W^R, respectively. As is well-known,

$$U^R \ll U , \quad Q^R \sim Q^K , \quad W^R \ll W ,$$

provided the characteristic transport velocity remains much less than the velocity of light. Then, under the influence of radiation,

$$Q = Q^K + Q^R \,, \tag{1.58}$$

Q^K and Q^R respectively denoting diffusion of heat flow by conduction and radiation.

1.3 RATE OF ENTROPY PRODUCTION

The entropy production discussed in Sections 1.2.2 and 1.2.3 is extended here to moving media which requires as well the consideration of the *momentum balance*. For the Stokesean fluid, this balance in terms of the usual nomenclature is

$$\rho \frac{Dv_i}{Dt} = -\frac{\partial p}{\partial x_i} + \frac{\partial \tau_{ij}}{\partial x_j} + \rho f_i \,. \tag{1.59}$$

In terms of the entropy flux,

$$\Psi_i = \frac{q_i}{T} \,, \tag{1.60}$$

the *entropy balance* (the Second Law balanced by the local entropy production) is

$$\rho \frac{Ds}{Dt} = -\frac{\partial \Psi_i}{\partial x_i} + s''' \,, \tag{1.61}$$

where s''' denotes the local entropy production. Also, the rate of the *conservation of total* (thermomechanical) *energy* (or the First Law) including the heat flux expressed in terms of the entropy flux,

$$\frac{\partial q_i}{\partial x_i} \equiv \frac{\partial}{\partial x_i}(\Psi_i T) = T\frac{\partial \Psi_i}{\partial x_i} + \Psi_i \frac{\partial T}{\partial x_i} \,, \tag{1.62}$$

is

$$\rho \frac{D}{Dt}\left(u + \frac{1}{2}v_i^2\right) = -\frac{\partial}{\partial x_i}(\Psi_i T) - \frac{\partial}{\partial x_i}(pv_i)$$
$$+ \frac{\partial}{\partial x_j}(\tau_{ij}v_i) + \rho f_i v_i + u''' \,, \tag{1.63}$$

u''' being the rate of energy generation by external (electromagnetic, chemical, nuclear) sources. Now, the fundamental difference of power,

Rate of total energy $-$ (Momentum)v_i $-$ (Rate of entropy)T ,

or,

Rate of (Total $-$ Mechanical $-$ Thermal) energy   (1.64)

leads, in terms of Eqs. (1.59), (1.61), (1.63) and the *conservation of mass*,

$$\frac{D\rho}{Dt} + \rho\frac{\partial v_i}{\partial x_i} = 0 , \qquad (1.65)$$

to

$$\rho\left(\frac{Du}{Dt} - T\frac{Ds}{Dt} + p\frac{Dv}{Dt}\right) = -\Psi_i\frac{\partial T}{\partial x_i} + \tau_{ij}s_{ij} + u''' - Ts''' , \quad (1.66)$$

where s_{ij} is the rate of deformation. For a reversible process, all forms of dissipation vanish, and

$$\left(\frac{Du}{Dt} - T\frac{Ds}{Dt} + p\frac{Dv}{Dt}\right) = 0 \qquad (1.67)$$

which is the Gibbs Thermodynamic relation. For an irreversible process assumed in *local equilibrium*, Eq. (1.67) continues to hold. Then, Eq. (1.66) gives the local entropy production

$$s''' = \frac{1}{T}\left[-\Psi_i\left(\frac{\partial T}{\partial x_i}\right) + \tau_{ij}s_{ij} + u'''\right] , \qquad (1.68)$$

where the terms in brackets respectively denote the dissipation of heat, work, electromagnetic, chemical and nuclear energy into internal energy. When radiation is appreciable, Ψ_i and q_i respectively denote the total entropy and heat fluxes involving the sum of the conductive flux and the radiative flux,

$$\Psi_i = \Psi_i^K + \Psi_i^R , \qquad q_i = q_i^K + q_i^R$$

and the local entropy production becomes

$$s''' = \frac{1}{T}\left[-\frac{1}{T}\left(q_i^K + q_i^R\right)\left(\frac{\partial T}{\partial x_i}\right) + \tau_{ij}s_{ij} + u'''\right] . \qquad (1.69)$$

Foregoing considerations will help us to understand dissipation in turbulent flows to be discussed in Chapter 4.

PROBLEMS

1-1) What are Oboukhov-Corrsin's time and velocity scales?

1-2) Why Oboukhov-Corrsin scales are for $Pr \to 0$.

1-3) Why the Oboukhov-Corrsin (thermal) scales are in terms of mechanical dissipation rather than thermal dissipation? *Hint:* Consider the role of Prandtl number in heat transfer.

1-4) Determine Batchelor's velocity and time scales by assuming similarity between velocity and temperature fields.

1-5) Why Batchelor scales are for $Pr \to \infty$?

1-6) Using the Gibbs relation, express the internal energy of the First Law,

$$dU = \delta Q + \delta \Phi \,,$$

in terms of Entropy, and divide each term by T. Explain the result.

1-7)

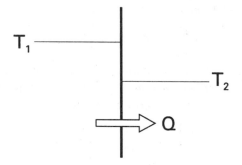

Fig. P1.1

Consider a reservoir at uniform temperature T_1 in contact with another at uniform temperature T_2 (Fig P1.1). In classical thermodynamics, the entropy difference between these reservoirs is given by

$$S_2 - S_1 = Q \left(\frac{1}{T_2} - \frac{1}{T_1} \right)$$

Interpret this relation in terms of the Second Law,

$$dS = \delta \Psi + \delta \Pi \,.$$

1-8) Reconsider the preceding problem in terms of an unsteady and a steady process shown in Fig. P1.2.

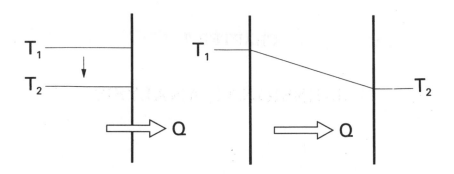

Fig. P1.2

1-9) Consider the Couette flow between two parallel plates which are kept at different temperatures.

   i) Write Momentum, Mechanical energy, the First Law and the Second Law for a differential control volume.

   ii) Express

$$\text{First Law} - \text{Mechanical Energy}$$

     in terms of Fourier's Law ad Newton's Law. Why conduction is in terms of a second derivative while the shear is in terms of the square of a first derivative?

   iii) Evaluate entropy production from

$$\text{First Law} - \text{Mechanical energy} - T \times \text{Second Law} \ .$$

1-10) Repeat the development in Section 1.3 on the rate of Entropy Production in terms of a suddenly pressurized stagnant fluid between tow parallel plates.

1-11) Repeat Section 1.3 in terms of a shock front. Use one-dimensional Cartesian coordinates.

1-12) Repeat Section 1.3 in terms of unsteady one-dimensional viscoelasticity. Use, for example, a beam subject to unsteady tension-compression stresses. What is the form of Gibbs relation for this case?

CHAPTER 2

DIMENSIONAL ANALYSIS

2.1 INTRODUCTION

When we have a complete understanding of physics and have no difficulty with formulation but are mathematically stuck before solution, dimensional analysis provides a functional (implicit) form of solution. Actually, there exists three distinct methods for dimensional analysis:

1) Formulation (nondimensionalized)

Whenever a formulation is readily available, term-by-term nondimensionalization of this formulation leads directly to the related dimensionless numbers. The procedure is not suitable to problems which cannot be readily formulated.

2) Π - Theorem[1]

If a formulation is not readily accessible but *all* physical and geometric quantities which characterize a physical situation are *clearly* known, we write an implicit relation among these quantities,

$$f(Q_1, Q_2, \ldots, Q_n) = 0 \,. \tag{2.1}$$

Expressing these quantities in terms of appropriate fundamental units, and making Eq. (2.1) independent of these fundamental units by an appropriate combination of Q's yields the dimensionless numbers.

---

[1] Dimensionless numbers obtained by this method are usually called Π's.

20

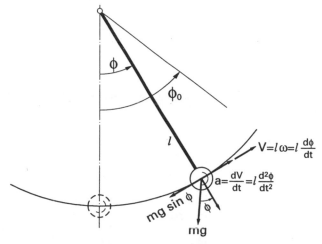

FIGURE 2.1: Simple pendulum.

3) Physical similitude

Ratios established from the individual terms of appropriate general principles gives the dimensionless numbers. The great convenience of this method is that there is no need to worry about an explicit formulation (required for the first method), except for a clear understanding of terms comprising a general principle. Also, there is no need to go through a nondimensionalization process (required for the second method) since a ratio between any two terms of a general principle is automatically dimensionless.

Let us illustrate the application of the foregoing methods in terms of an illustrative example based on a simple pendulum (in vacuum). We wish to determine the period of this pendulum by dimensional analysis.

From the tangential component of Newton's Law of Motion, we have the governing equation (Fig. 2.1),

$$\frac{d^2\phi}{dt^2} + \frac{g}{\ell}\sin\phi = 0 \ . \tag{2.2}$$

For the *first method*, we have from the nondimensionalization of Eq. (2.2) in terms of period T,

$$\frac{d^2\phi}{d(t/T)^2} + \left(T^2\frac{g}{\ell}\right)\sin\phi = 0$$

which suggests the functional (implicit) relationship

$$\phi = f\left(\frac{t}{T}, T^2\frac{g}{\ell}\right) . \tag{2.3}$$

However, we are not interested in instantaneous position ϕ (of the pendulum) but rather its extrema ϕ_o for which t/T assumes integer values, $1, 2, 3, \ldots$. Consequently,

$$\phi_o = f\left(T^2\frac{g}{\ell}\right) .$$

Inverting this functional relationship, and expressing the result in terms of the period rather than its square, we have

$$T\sqrt{\frac{g}{\ell}} = f(\phi_o) . \tag{2.4}$$

For the *second method* (the Π-Theorem), recalling the fact that the tangential momentum is balanced by the tangential component of the gravitational body force, and from the inspection of this balance, we conclude

$$T = f(m, g, \ell, \phi_o) , \tag{2.5}$$

where m, g, ℓ and ϕ_o all are independent quantities. In terms of three fundamental units of mechanics M, L, T,[2] Eq. (2.5) may be expressed as

$$[T] \equiv f\left[M, \frac{L}{T^2}, L, 0\right] . \tag{2.6}$$

Now, we begin rearranging Eq. (2.5) in such a way that, with each arrangement, it becomes independent of one fundamental unit. First, the dimensional homogeneity in M suggests

$$T = f_1(g, \ell, \phi_o) ,$$

or, in terms of the fundamental units,

$$[T] \equiv f_1\left[\frac{L}{T^2}, L, 0\right] .$$

---

[2]or F, L, T.

Eliminating L, for example, by ratio g/ℓ, yields

$$T = f_2\left(\frac{g}{\ell}, \phi_o\right) ,$$

or, in terms of the fundamental units,

$$[T] \equiv f_2\left[\frac{1}{T^2}, 0\right] .$$

Finally, eliminating T^3 by product $T\sqrt{\frac{g}{\ell}}$, gives dimensionless relation

$$T\sqrt{\frac{g}{\ell}} = f_3(\phi_o)$$

which is identical to Eq. (2.4). Note that the number of steps in the foregoing nondimensionalization procedure is equal to the number of fundamental units. Consequently, the number of dimensionless numbers is equal to the difference between the number of dimensional quantities in the original statement of a problem and the number of fundamental units. That is, Eq. (2.5) is in terms of 5 quantities, and since there are 3 fundamental units, the result involves $5 - 3 = 2$ dimensionless numbers.

For the *third method* (the Physical Similitude), consider the tangential balance between the inertial and gravitational forces, $F_I \sim F_G$, or the ratio

$$F_I/F_G \sim 1 . \tag{2.7}$$

An order of magnitude interpretation of this ratio in terms of ϕ_o and T reveals

$$\frac{m\ell\phi_o/T^2}{mgF(\phi_o)} \sim 1 ,$$

$F(\phi_o)$ indicating the tangential component of the gravitational force as a function of ϕ_o, or

$$T^2\frac{g}{\ell} \sim \frac{\phi_o}{F(\phi_o)} ,$$

or

$$T\sqrt{\frac{g}{\ell}} = f(\phi_o) . \tag{2.4}$$

---

[3]Note the use of T for period as well as a fundamental unit.

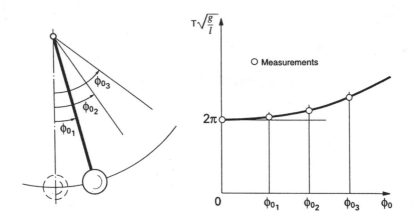

FIGURE 2.2: Experiments with simple pendulum.

Thus, by three distinct methods we are able to show that the dimension-less period of a simple pendulum in vacuum depends only on its initial displacement. Now, combining Eq. (2.4) with a simple experiment to be performed by one pendulum with a number of ϕ_o's (Fig. 2.2), we can determine the explicit form of Eq. (2.4).

As is well known, for small displacements from the equilibrium, assuming $\sin \phi \cong \phi$, Eq. (2.2) is reduced to

$$\frac{d^2\phi}{dt^2} + \frac{g}{\ell}\phi = 0 \tag{2.8}$$

which characterizes a harmonic motion with angular frequency $\omega = \sqrt{g/\ell}$. Consequently,

$$T = \frac{2\pi}{\omega} = 2\pi\sqrt{\frac{\ell}{g}} \tag{2.9}$$

which turns out to be independent of ϕ_o because of the assumed small oscillations.

Having learned the foundations of dimensional analysis in terms of an illustrative example, we proceed now to examples more relevant to our convection studies. Consider first an isothermal flow problem which will be useful for enthalpy flow of convection problems.

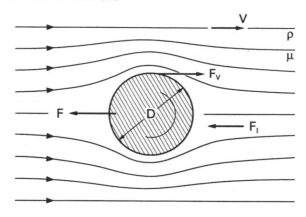

FIGURE 2.3: Forced flow over a sphere.

2.2 A FORCED FLOW

Let a solid sphere of diameter D be immersed and held in an incompressible fluid steadily streaming with uniform velocity V (Fig. 2.3). The density and viscosity of the fluid are ρ and μ, respectively. We wish to determine the drag force F on the sphere.

Since the differential formulation of a viscous flow near a sphere is not the concern of this monograph, we proceed with the Π-Theorem. In view of the fact that the drag force is balanced by the inertial and viscous forces, we assume

$$F = f(V, D, \rho, \mu) \qquad (2.10)$$

which may be expressed in terms of the fundamental units as

$$\left[\frac{ML}{T^2}\right] \equiv f\left[\frac{L}{T}, L, \frac{M}{L^3}, \frac{M}{LT}\right].$$

Now, we begin rearranging Eq. (2.10) by making it independent of one fundamental unit at a time. Since the mass-dependence is the simplest one, we begin with mass. To eliminate M, we pick any one of the mass dependent terms of the right-hand side, ρ or μ. For example, we pick μ (later we comment on what would happen if we would have picked ρ instead) and combine it with F and ρ in such a way that the mass-dependence disappears. Thus

$$\frac{F}{\mu} = f_1\left(V, D, \frac{\rho}{\mu}\right) \qquad (2.11)$$

which may be expressed in terms of the remaining fundamental units as

$$\left[\frac{L^2}{T}\right] \equiv f_1\left[\frac{L}{T}, L, \frac{T}{L^2}\right] \ .$$

Clearly, the time-dependence of Eq. (2.11) is simpler than its length dependence. To eliminate T, we pick any one of the time-dependent terms on the right-hand side, V or ρ/μ. Since the final dimensionless numbers will ultimately involve all quantities describing Eq. (2.10), and since we already manipulated with μ, this time we pick V and combine it with F/μ and ρ/μ in such a way that the time-dependence disappears. Thus

$$\frac{F}{\mu V} = f_2\left(D, \frac{\rho V}{\mu}\right) \tag{2.12}$$

which, in terms of the length unit, may be expressed as

$$[L] \equiv f_2\left[L, \frac{1}{L}\right] \ .$$

Finally, eliminating L by combining D with other terms of Eq. (2.12), we get

$$\frac{F}{\mu V D} = f_3\left(\frac{\rho V D}{\mu}\right) = f_3(Re) \ . \tag{2.13}$$

Now, let us go back to Eq. (2.10) and this time we make this equation independent of M by manipulating the mass-dependent terms with ρ rather than μ. This leads to

$$\frac{F}{\rho V^2 D^2} = f_3\left(\frac{\mu}{\rho V D}\right) = f_4(Re) \ . \tag{2.14}$$

Since the dimensional analysis can provide only a functional (implicit) relationship between dimensionless numbers, Eqs. (2.13) and (2.14) are synonymous dimensionless results. Furthermore, since Eq. (2.14) is a functional relationship, it can be rearranged as

$$\frac{F}{\rho V^2 D^2} Re = f_5(Re)$$

which is identical to Eq. (2.13). That is, by suitable transformations, a dimensionless result can be made identical to another dimensionless result.

Dimensional analysis offers no clue as to which one of Eqs. (2.13) and (2.14) may be most convenient. Aside from the obvious fact that dependent variable F should be included in only one dimensionless number, it is necessary to rely on past experience and physical insight in the selection of one of these relations.

Next, we proceed to dimensional analysis of the same problem by the method of physical similitude. Since force F on the sphere exerted by the moving fluid is balanced with the inertial and viscous forces,

$$F = f(F_I, F_V)$$

from which we may establish ratios

$$\frac{F}{F_V} \sim \frac{F}{D^2 \mu(V/D)} = \frac{F}{\mu V D} ,\qquad (2.15)$$

$$\frac{F_I}{F_V} \sim \frac{\rho D^3(V^2/D)}{D^2 \mu(V/D)} = \frac{\rho V D}{\mu} = Re ,\qquad (2.16)$$

and

$$\frac{F}{F_I} \sim \frac{F}{\rho D^3(V^2/D)} = \frac{F}{\rho V^2 D^2} .\qquad (2.17)$$

Clearly, from $F_I = m(dV/dt)$ we have $F_I \sim mV/t$ and (in view of $m \sim \rho D^3$ and $t \sim D/V$) get $F_I \sim \rho D^3(V^2/D)$. Also, for a Newtonian fluid, from $\tau = \mu(dU/dy)$ we have $F_V \sim D^2\mu(V/D)$. Eqs. (2.15) and (2.16) are the dimensionless numbers associated with Eq. (2.13), and Eqs. (2.16) and (2.17) are those associated with Eq. (2.14).

The concept of physical similitude becomes also quite useful for experiments to be conducted with scaled models rather than the actual prototype. Physical similitude is said to exist between two systems if corresponding dimensionless numbers have the same value. Geometric similitude is a prerequisite for the physical similitude. Further elaborations on the similitude, however, belong to texts on fluid mechanics.

Having learned the dimensionless numbers associated with forced flows, we proceed next to the dimensionless numbers associated with buoyancy driven flows.

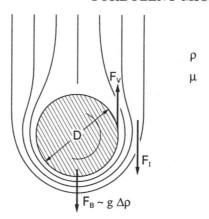

FIGURE 2.4: Free fall of a sphere.

2.3　A FREE FALL

Consider the solid sphere of the preceding example. Let the sphere now fall, under the effect of gravity, in a fluid of density ρ and viscosity μ (Fig. 2.4). The difference between the density of the sphere and that of the fluid is $\Delta\rho$. We wish to determine the terminal velocity of the sphere.

In a manner similar to the preceding problem, we begin with the Π-Theorem. First, replacing F of Eq. (2.10) with the buoyant force per unit volume $g\Delta\rho$, we write

$$g\Delta\rho = f(V, D, \rho, \mu) \ ,$$

and, because V now is the dependent variable, rearrange for V,

$$V = f(g\Delta\rho, D, \rho, \mu) \ . \tag{2.18}$$

In terms of the fundamental units, Eq. (2.28) is equivalent to

$$\left[\frac{L}{T}\right] \equiv f\left[\frac{M}{L^2T^2}, L, \frac{M}{L^3}, \frac{M}{LT}\right] \ .$$

Next, we begin rearranging Eq. (2.18) so that it becomes suitable to the elimination of one fundamental unit at a time. To eliminate M, we pick ρ, for example, and combine it with $g\Delta\rho$ and μ in such a way that the mass dependence disappears. Thus

$$V = f_1\left(g\frac{\Delta\rho}{\rho}, D, \frac{\mu}{\rho}\right) \tag{2.19}$$

which, in terms of the remaining fundamental units, is equivalent to

$$\left[\frac{L}{T}\right] \equiv f_1 \left[\frac{L}{T^2}, L, \frac{L^2}{T}\right] \, .$$

To eliminate T, we pick $\mu/\rho = \nu$, for example, and combine it with V and $g(\Delta\rho/\rho)$ in such a way that the time-dependence disappears. Thus

$$\frac{\rho V}{\mu} = f_2 \left(\frac{g}{\nu^2}(\frac{\Delta\rho}{\rho}), D\right) \tag{2.20}$$

which, in terms of the length unit, is equivalent to

$$\left[\frac{1}{L}\right] \equiv f_2 \left[\frac{1}{L^3}, L\right] \, .$$

Finally, eliminating L by combining D with other terms of Eq. (2.20), we get

$$\frac{\rho V D}{\mu} = f_3 \left(\frac{g}{\nu^2}(\frac{\Delta\rho}{\rho})D^3\right) \, , \tag{2.21}$$

where

$$\frac{g}{\nu^2}\left(\frac{\Delta\rho}{\rho}\right) D^3 = Gr \tag{2.22}$$

is the *Grashof number*. Thus, the terminal velocity of a buoyancy driven body is found to be governed by dimensionless relation

$$Re = f(Gr) \, . \tag{2.23}$$

Next, we proceed to dimensional analysis of the same problem by the method of physical similitude. For the sphere, the buoyant force is balanced by the viscous force, and for the fluid, the viscous force is balanced by the inertial force,

$$
\begin{array}{c|c}
\text{Sphere} & \text{Fluid} \\
F_B \sim & F_V \quad \sim F_I \\
\end{array}
$$

which lead to the following force ratios

$$\frac{F_I}{F_V} \sim \frac{\rho V D}{\mu} = Re \, , \tag{2.16}$$

$$\frac{F_B}{F_V} \sim \frac{g\Delta\rho D^3}{\mu(V/D)D^2} = \frac{g\Delta\rho D^2}{\mu V} . \tag{2.24}$$

Now, we refer to the general fact that the physics of any problem may be described by one dimensionless number depending on other dimensionless number(s) composed only of independent quantities. For the present problem

$$\text{Velocity} = f(\text{ Buoyancy }) .$$

Velocity is nondimensionalized with Re, and with the following combination, obtained from the product of Eqs. (2.16) and (2.24), buoyancy is nondimensionalized with

$$\frac{F_B}{F_V} \times \frac{F_I}{F_V} \sim \frac{g\Delta\rho D^2}{\mu V} \times \frac{\rho V D}{\mu} = \frac{g}{\nu^2}\left(\frac{\Delta\rho}{\rho}\right)D^3 = Gr$$

which is independent of velocity. Thus the physical similitude leads us to a relation among force ratios

$$\frac{F_I}{F_V} = f\left(\frac{F_B}{F_V} \times \frac{F_I}{F_V}\right) \tag{2.25}$$

which is identical to Eq. (2.23), the result already obtained by employing the Π-Theorem.

Now we are ready for a dimensional analysis of convection problems, and begin with forced convection because of its relative simplicity.

2.4   FORCED CONVECTION

As is well known, the Nusselt number is the dimensionless wall gradient of fluid temperature. Ignoring the method of nondimensionalized governing equations because of its complexity, we proceed to the Π-Theorem.

Consider, for example, a steady two-dimensional problem with temperature distribution

$$\frac{T_w - T}{T_w - T_\infty} = f(x, y, \underbrace{\frac{Q_H}{(V, \rho, \mu)\, c_p}}, \underset{k}{Q_K})$$

or,

$$\frac{T_w - T}{T_w - T_\infty} = f(x, y, V, \rho, \mu, c_p, k) ,$$

T_w, T, T_∞ respectively being the wall temperature, fluid temperature and its free stream value. The wall gradient of the fluid temperature gives the *local heat transfer coefficient*,

$$h_x = f(x, V, \rho, \mu, c_p, k)$$

whose average over distance D (or ℓ) gives the (*average*) *heat transfer coefficient*,

$$h = f(D, V, \rho, \mu, c_p, k) , \tag{2.26}$$

where the right-hand side composed only of independent quantities. For thermal problems, a fourth fundamental unit is needed in addition to the three fundamental units of mechanics. This unit is usually assumed to be temperature θ[4]. In terms of the four fundamental units, Eq. (2.26) may be expressed as

$$\left[\frac{M}{T^3\theta}\right] \equiv f\left[L, \frac{L}{T}, \frac{M}{L^3}, \frac{M}{LT}, \frac{L^2}{T^2\theta}, \frac{ML}{T^3\theta}\right] .$$

(For the units of c_p, recall the definition of stagnation enthalpy[5] $\hat{h}^o = \hat{h} + V^2/2 + gz$ and use $\hat{h} \sim c_p T \sim V^2/2$ which gives $c_p\theta \sim L^2/T^2$; for the units of h, use $q \sim$ Power/Area $=$ Force$\times$ Velocity/Area which yields $h\theta \sim (ML/T^2)(L/T)/L^2$, and for the units of k, note $hL \sim k$). We proceed now to dimensionless numbers associated with forced convection by successively eliminating the fundamental units from Eq. (2.26).

Again we begin the elimination process by M. Combining ρ, for example, with other mass-dependent quantities, we have from Eq. (2.26)

$$\frac{h}{\rho} = f_1\left(D, V, \nu, c_p, \frac{k}{\rho}\right) , \tag{2.27}$$

where $\nu = \mu/\rho$ is the kinematic viscosity (or momentum diffusivity). In terms of the fundamental units, Eq. (2.27) is equivalent to

$$\left[\frac{L^3}{T^3\theta}\right] \equiv f_1\left[L, \frac{L}{T}, \frac{L^2}{T}, \frac{L^2}{T^2\theta}, \frac{L^4}{T^3\theta}\right] .$$

---

[4] Or heat flux Q

[5] $\hat{h}$ is used for enthalpy to distinguish from heat transfer coefficient h.

Now the temperature dependence appears to be the simplest one. Elimination of θ from Eq. (2.27) by combining c_p, for example, with other temperature-dependent quantities yields

$$\frac{h}{\rho c_p} = f_2(D, V, \nu, \alpha) , \qquad (2.28)$$

where $\alpha = k/\rho c_p$ is the thermal diffusivity. Eq. (2.28) is equivalent to

$$\left[\frac{L}{T}\right] \equiv f_2\left[L, \frac{L}{T}, \frac{L^2}{T}, \frac{L^2}{T}\right] .$$

Time-dependence of Eq. (2.28) is somewhat easier than its length dependence. Eliminating T by combining V, for example, with other time dependent quantities[6] gives

$$\frac{h}{\rho c_p V} = f_3\left(D, \frac{V}{\nu}, \frac{V}{\alpha}\right) , \qquad (2.29)$$

which is equivalent to

$$[0] \equiv f_3\left[L, \frac{1}{L}, \frac{1}{L}\right] ,$$

where 0 corresponding to a dimensionless quantity. Finally, eliminating L, we have

$$\frac{h}{\rho c_p V} = f_4\left(\frac{VD}{\nu}, \frac{VD}{\alpha}\right) ,$$

or,

$$St = f(Re, Pe) , \qquad (2.30)$$

where $St = h/\rho c_p V$ is the *Stanton number* and $Pe = VD/\alpha$ is the *Peclet number* (or Thermal Reynolds number). Noting that

$$Pe = RePr ,$$

$$St = Nu/Pe = Nu/RePr$$

Eq. (2.30) may be written alternatively as

$$\boxed{Nu = f(Re, Pr)} , \qquad (2.31)$$

---

[6]T may also be eliminated with either α or ν.

where Re is the only dimensionless number depending on velocity. This is the form used most frequently.

Next, we proceed to a dimensional analysis of the same problem by the method of physical similitude. From the definition of

$$\text{Convection} = \text{Conduction in } \underbrace{\text{moving media}}_{\underbrace{\text{Inertial, Viscous forces}}_{F_I, F_V}} \quad ,$$

we have

$$Q_C = f(F_I, F_V, Q_H, Q_K) \qquad (2.32)$$

where Q_C denotes convection, Q_H enthalpy flow, Q_K conduction, F_I inertial force and F_V viscous force. Next, we establish the ratios

$$\frac{Q_C}{Q_K} \sim \frac{hD^2\theta}{kD^2(\theta/D)} = \frac{hD}{k} = Nu \, , \qquad (2.33)$$

$$\frac{F_I}{F_V} \sim \frac{VD}{\nu} = Re \, , \qquad (2.16)$$

$$\frac{Q_H}{Q_K} \sim \frac{\rho c_p V D^2 \theta}{kD^2(\theta/D)} = \frac{VD}{\alpha} = Pe \, , \qquad (2.34)$$

where Nu is the heat transfer coefficient nondimensionalized relative to conduction, and Re and Pe are obtained from the nondimensionalization of momentum and thermal energy, respectively. For an incompressible flow, momentum is decoupled from energy, and is characterized by Re. However, energy is coupled to momentum through enthalpy flow, and Pe number is silent to this coupling. Eliminating the velocity between Eqs. (2.16) and (2.34) we obtain

$$\frac{Q_H}{Q_K} \times \frac{F_V}{F_I} \sim \frac{\nu}{\alpha} = Pr \, , \qquad (2.35)$$

which describes the coupling of energy to momentum. Also, Pr characterizes the diffusion of momentum relative to that of heat, and is the only dimensionless number in terms of physical properties.

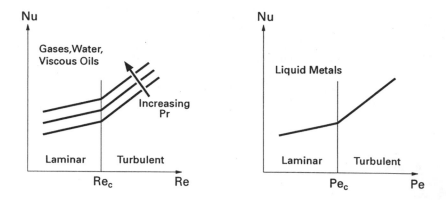

FIGURE 2.5: Correlation of forced convection data on (a) gases, water, and viscous oils,
(b) liquid metals.

Experimental data associated with gases, water and viscous oils may be correlated with Eq. (2.31) as shown in Fig. 2.5a, where Re_c denotes the critical Reynolds number at which the laminar flow is unstable. Beyond Re_c the forced convection becomes eventually turbulent. Eq. (2.31) does not correlate the liquid metal data. For liquid metals, viscous forces are small and the momentum equation degenerates to a limit of uniform velocity, and the importance of the Reynolds number diminishes. Consequently, as shown in Fig. 2.5b,

$$\boxed{Nu = f(Pe)} \qquad (2.36)$$

correlates the data on liquid metals.

Having learned the dimensionless numbers associated with forced convection we proceed now to those for natural convection.

2.5   NATURAL CONVECTION

So far we learned that the correlation of forced convection begins with

$$h = f(D, V, \rho, \mu, c_p, k) , \qquad (2.26)$$

where the right-hand terms are made only of independent quantities. Since the velocity of natural convection depends on buoyancy, we may utilize Eq. (2.26) for natural convection after replacing V with a buoyancy term.

Buoyancy force/unit volume is

$$g\Delta\rho .$$

For small temperature differences, assume

$$\Delta\rho = \left(\frac{d\rho}{dT}\right)\Delta T$$

which may be rearranged in terms of the coefficient of thermal expansion, $\beta = -(1/\rho)(d\rho/dT)$, as

$$\Delta\rho \sim \beta\rho\Delta T .$$

Consequently, the buoyancy force/volume becomes

$$g\beta\rho\Delta T .$$

Noting that Eq. (2.26) already involves ρ because of inertial force, and now replacing V of this equation with $g\beta\Delta T$, we have

$$h = f(D, g\beta\Delta T, \rho, \mu, c_p, k) . \tag{2.37}$$

Consider first the method of Π-Theorem, and accordingly, express Eq. (2.37) in terms of the four fundamental units as

$$\left[\frac{M}{T^3\theta}\right] \equiv f\left[L, \frac{L}{T^2}, \frac{M}{L^3}, \frac{M}{LT}, \frac{L^2}{T^2\theta}, \frac{ML}{T^3\theta}\right]$$

and begin the elimination process by the mass-dependent terms. We already used μ (in connection with flow around a sphere) and ρ in connection with forced convection) for this elimination. Let us see what happens when we eliminate M by combining k with other mass-dependent quantities:

$$\frac{h}{k} = f_1\left(D, g\beta\Delta T, \frac{\rho}{k}, \frac{\mu}{k}, c_p\right) \tag{2.38}$$

which, in terms of fundamental units, is equivalent to

$$\left[\frac{1}{L}\right] \equiv f_1\left[L, \frac{L}{T^2}, \frac{T^3\theta}{L^4}, \frac{T^2\theta}{L^2}, \frac{L^2}{T^2\theta}\right] .$$

At this stage, the temperature dependence appears to be the simplest one. Elimination of θ from Eq. (2.38) by combining c_p, for example, with other temperature-dependent quantities yields

$$\frac{h}{k} = f_2 (D, g\beta\Delta T, \alpha, Pr) , \qquad (2.39)$$

where $\alpha = k/\rho c_p$ is the thermal diffusivity and $Pr = \mu c_p/k = \nu/\alpha$ is the Prandtl number. Eq. (2.39) is equivalent to

$$\left[\frac{1}{L}\right] \equiv f_2 \left[L, \frac{L}{T^2}, \frac{L^2}{T}, 0\right] .$$

Eliminating the time dependence from Eq. (2.39) by combining, for example, α with $g\beta\Delta T$ we get

$$\frac{h}{k} = f_3 \left(D, \frac{g\beta\Delta T}{\alpha^2}, Pr\right) \qquad (2.40)$$

which is equivalent to

$$\left[\frac{1}{L}\right] \equiv f_3 \left[L, \frac{1}{L^3}, 0\right] .$$

Finally, eliminating L yields

$$\frac{hD}{k} = f_4 \left(\frac{g\beta\Delta T D^3}{\alpha^2}, Pr\right) ,$$

or,

$$Nu = f(\Pi, Pr) , \qquad (2.41)$$

where $\Pi = g\beta\Delta T D^3/\alpha^2$ is a dimensionless number. Noting that

$$\frac{\Pi}{Pr} = \frac{g\beta\Delta T D^3}{\alpha\nu} = Ra ,$$

Eq. (2.41) may be rearranged in terms of Ra as

$$\boxed{Nu = f(Ra, Pr)} , \qquad (2.42)$$

where Ra is the *Rayleigh number*.

Next, we proceed to dimensional analysis of the same problem by the method of physical similitude. From the definition of

Natural Convection = Conduction in $\underbrace{\text{moving media}}$

$$\underbrace{\text{Buoyant, Inertial, Viscous forces}}$$

$$F_B, F_I, F_V$$

we have

$$Q_C = f(F_B, F_I, F_V, Q_H, Q_K) , \qquad (2.43)$$

F_B denoting buoyant force, others already defined following Eq. (2.32). Next, we establish the following ratios

$$\frac{Q_C}{Q_K} = \frac{hD}{k} = Nu , \qquad (2.33)$$

$$\frac{F_I}{F_V} = \frac{VD}{\nu} = Re , \qquad (2.16)$$

$$\frac{F_B}{F_I} \sim \frac{mg\beta\Delta T}{mV^2/D} = \frac{g\beta\Delta T D}{V^2} , \qquad (2.44)$$

$$\frac{Q_H}{Q_K} = \frac{VD}{\alpha} = Pe . \qquad (2.34)$$

Here, we recall the fact that dimensionless numbers are composed only of independent quantities, and note that V is *not* an independent quantity for natural convection. Consequently, Eqs. (2.16), (2.34) and (2.44) may describe natural convection after they are combined and made independent of V. For example, the following combination of Eqs. (2.34) and (2.16)

$$\frac{F_B}{F_V} \times \frac{Q_H}{Q_K} = \frac{(g\beta\Delta T)D^2}{\nu V} \times \frac{VD}{\alpha} = \frac{g}{\nu\alpha}(\beta\Delta T) D^3 = Ra \qquad (2.45)$$

which is independent of velocity, shows how the momentum is coupled to thermal energy through its buoyancy. Also, the following combination of Eqs. (2.16) and (2.34)

$$\frac{Q_H}{Q_K} \times \frac{F_V}{F_I} = \frac{VD}{\alpha}\frac{\nu}{VD} = \frac{\nu}{\alpha} = Pr \qquad (2.35)$$

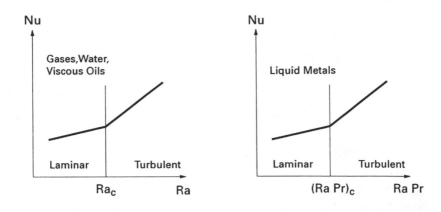

FIGURE 2.6: Correlation of natural convection data on (a) gases, water, and viscous oils, (b) liquid metals.

which is also independent of velocity, shows how the thermal energy is coupled to momentum through its enthalpy flow.

For $Pr > 1$, inertial effect is negligible and Eq. (2.42) is reduced to

$$Nu = f(Ra) \ . \tag{2.46}$$

Accordingly, the experimental data on gases, water and viscous oils are correlated with Eq. (2.46) as shown in Fig. 2.6a.

For $Pr \ll 1$, viscous effect is negligible, and Eq. (2.42) is reduced to

$$Nu = f(RaPr) \ . \tag{2.47}$$

Experimental data on liquid metals are correlated with Eq. (2.47) as shown in Fig. 2.6b.

It is important to note here the conceptual difference between the Reynolds number of forced convection and the Rayleigh number of natural convection. Re results from the nondimensionalized momentum (of forced convection) which is decoupled from thermal energy of incompressible (and constant property) fluids. On the other hand, Ra characterizes the coupling (through buoyancy) of momentum to energy.[7]

---

[7]Grashof number, $Gr = g\beta\Delta T D^3/\nu^2$ results from the nondimensionalized momentum only, it alone cannot characterize and should not be used for natural convection related to a temperature difference.

Actually, the foregoing arguments can be extended one step further by recognizing the fact that buoyant force drives both inertial and viscous forces, and

$$F_B \sim F_I + F_V . \tag{2.48}$$

Then, Eq. (2.43) becomes

$$Q_C = f(F_B, F_I + F_V, Q_H, Q_K) . \tag{2.49}$$

Now, coupling

$$\frac{F_B}{F_I + F_V} = \frac{F_B/F_V}{1 + F_I/F_V} , \tag{2.50}$$

with Eq. (2.34) in the following manner, eliminates velocity and leads to

$$\Pi_N = \frac{\left(\dfrac{F_B}{F_V}\right)\dfrac{Q_H}{Q_K}}{1 + \left(\dfrac{F_I}{F_V}\right)\dfrac{Q_K}{Q_H}} \sim \frac{\dfrac{g\beta\Delta T D^2}{\nu V} \times \dfrac{VD}{\alpha}}{1 + \dfrac{VD}{\nu} \times \dfrac{\alpha}{VD}} \tag{2.51}$$

or,

$$\Pi_N \sim \frac{Ra}{1 + Pr^{-1}} = \frac{RaPr}{1 + Pr} . \tag{2.52}$$

Although the existence of Π_N has never been recognized, an early integral solution to laminar natural convection given by Squire (1938) can be expressed in terms of Π_N (see Prob. 2.1). The limits of Eq. (2.52) for $Pr \to \infty$ and $Pr \to 0$ (explicit in the first and second equality, respectively) are

$$\lim_{Pr\to\infty} \Pi_N = Ra , \tag{2.53}$$

$$\lim_{Pr\to 0} \Pi_N = RaPr . \tag{2.54}$$

Also, in terms of Π_N, Eq. (2.42) becomes

$$Nu = f(\Pi_N) \tag{2.55}$$

whose limits for $Pr \to \infty$ and $Pr \to 0$ respectively are Eqs. (2.46) and (2.47), as expected.

Having learned the functional (implicit) relation among the dimensionless numbers of forced convection and among the dimensionless numbers of natural convection, we now proceed to next chapter for explicit relations among these numbers for turbulent flows.

PROBLEMS

2-1) Assuming a parabolic temperature and cubic velocity for natural convection near a vertical plate at temperature T_w in an ambient at temperature T_∞, Squire (1938) obtained

$$Nu = 0.508 Pr^{1/2}(Pr + 0.952)^{-1/4}Gr^{1/4} .$$

Show, by rearranging this equation in terms of Π_N, that

$$Nu = 0.508\Pi_N^{1/4} , \quad \Pi_N = \frac{Ra}{1 + 0.952/Pr} .$$

2-2) Repeat Prob. 2-1 for the limit $Pr \to 0$ and $Pr \to \infty$.

2-3) Consider a flow over a deformable sphere. Including the effect of gravity g and surface tension σ (force/length), assume

$$F = f(V, D, \rho, \mu, g, \sigma) .$$

Show by the Π-Theorem and the Physical Similitude that the drag force may be nondimensionalized to give

$$\frac{F}{\rho V^2 D^2} = f(Re, Fr, We) ,$$

where $Fr = V/\sqrt{gD}$ is the *Froude number*, and $We = \rho V^2 D/\sigma$ is the *Weber number*. Explain clearly the physics characterized by these numbers.

2-4) Including the effect of surface tension σ to Eq. (2.18), extend the problem stated in Section 2.3 to a gas bubble rising in a liquid. For the terminal velocity of the bubble, assume

$$V = f(g\Delta\rho, D, \rho, \mu, \sigma) .$$

Show by the Π-Theorem and the Physical Similitude that the velocity may be nondimensionalized to give

$$Re = f(Gr, Bo) ,$$

where $Bo = g\Delta\rho D^2/\sigma$ is the *Bond number*.

2-5) Express in terms of appropriate dimensionless numbers the diameter of droplets formed by a liquid discharging with a specified velocity (or under the effect of a pressure gradient) from a horizontal tube.

2-6) Express in terms of appropriate dimensionless numbers the diameter of droplets formed by a liquid discharging under the effect of gravity from a vertical tube.

2-7) Consider the flow between two co-axial cylinders in relative rotation. Write a dimension less relation between torque and angular frequency.

2-8) A wooden sphere is held by a string in a water stream. Determine the string force by means of a dimensional analysis.

2-9) What are the dimensionless numbers of combined forced-natural convection?

2-10) Consider the natural convection from a horizontal cylinder rotating with angular frequency ω. Peripheral surface temperature of the cylinder is T_w and the ambient temperature is T_∞.

Assuming the convection resulting from rotation and that from gravity can be superimposed, express the Nusselt number in terms of appropriate dimensionless numbers.

2-11) Discuss the physics of a hot air balloon in terms of appropriate dimensionless numbers.

CHAPTER 3

TURBULENCE

Turbulence results from the unstable growth of infinitesimal disturbances existing in a flow. Studies on its origin begin with the small
disturbance theory of instability which we consider here in terms of a
number of illustrative examples.

The theory considers the change of total energy of a system resulting
from an infinitesimal disturbance introduced to the initial state of the
system. Let this change of total energy be ΔE. The stability criterion
is then

$$\Delta E > 0 , \quad \text{stable} ,$$

$$\Delta E = 0 , \quad \text{marginal} ,$$

$$\Delta E < 0 , \quad \text{unstable} .$$

The celebrated example is the change of the potential energy of a marble
corresponding to its infinitesimal displacement in a bowl or over an
inverted bowl, or on a flat horizontal plate. Fig. 3.1 shows the stable
and unstable cases, and the marginal state separating these cases.

Two examples, one from mechanics and one from conduction, elaborate the application of the energy method to instability problems. The
first example, depicted in Fig. 3.2, deals with the buckling load P of a
rigid vertical beam, hinged at the bottom and attached to a horizontal
spring at the top. The beam length is ℓ, the spring constant is k.

For a small disturbance α, the decrease in potential energy of the
load,

$$-P\ell(1 - \cos \alpha) \cong -\frac{P\ell\alpha^2}{2} ,$$

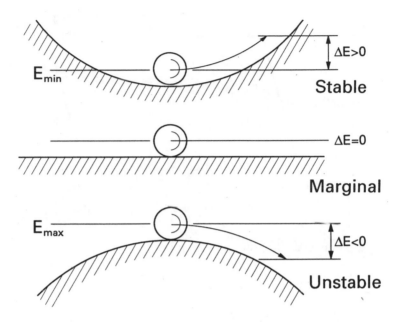

FIGURE 3.1: Original stability problem.

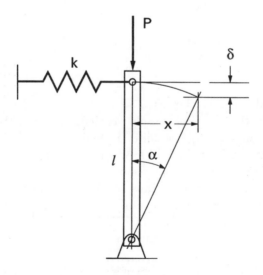

FIGURE 3.2: Stability of a rigid beam.

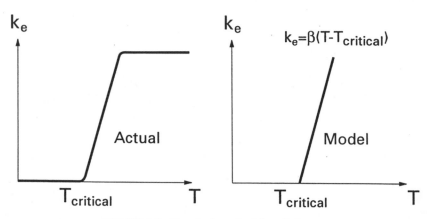

FIGURE 3.3: Electrical conductivity of glass.

and the increase in potential energy of the spring,

$$\frac{kx^2}{2} \simeq \frac{k\ell^2\alpha^2}{2} \, ,$$

give the change in total potential energy of the system,

$$\Delta E = \frac{k\ell^2\alpha^2}{2} - \frac{P\ell\alpha^2}{2} = 0 \, ,$$

yielding the marginal force

$$P = k\ell \, ,$$

above which the beam collapses.

In terms of the foregoing simple problem, we learn the application of the energy method to instability in rigidbody mechanics. Before addressing the instability in fluid mechanics problems, we now consider a transitory problem from conduction.

Consider a flat glass plate of thickness ℓ. The plate is heated electrically. The temperature dependence of the electrical conductivity $k_e(T)$ of glass and its model are given in Fig. 3.3. Plate walls are kept at a constant temperature. The thermal instability (buckling) of the plate is required.

Since the electrical conductivity of glass below a critical temperature is very small, the energy generation $u''' = e^2 k_e$ below this temperature is negligible. Assume that the temperature distribution in the plate stays above and the walls at the critical temperature T_{crit}.

The change in thermal energy,

$$\Delta E = \text{thermal energy [loss(by conduction)} - \text{generation]} = 0,$$
$$(3.1)$$

gives the stability criterion,

$$\Delta E = -k \left(\frac{d\theta}{dx}\right)\Bigg|_{x=0}^{x=\ell} - e^2\beta \int_0^\ell \theta \, dx = 0 \,, \qquad (3.2)$$

or

$$-k \left(\frac{d\theta}{dx}\right)\Bigg|_{x=0}^{x=\ell} = e^2\beta \int_0^\ell \theta \, dx \,, \qquad (3.3)$$

$\theta = T - T_{\text{crit}}$ and e being the electric potential. The energy method applied to a spatially distributed problem requires the selection of an approximate profile for the dependent variable. For the present problem, we need to know the temperature distribution. Assuming, for example, that

$$\theta = A \sin \pi \left(\frac{x}{\ell}\right) \,, \qquad (3.4)$$

A being an arbitrary small amplitude, and inserting this profile into Eq. (3.3) results in

$$e \left(\frac{\beta}{k}\right)^{1/2} = \frac{\pi}{\ell} \qquad (3.5)$$

for the marginal electric potential above which the system is unstable and the temperature of the glass plate increases until melt-down.

Actually, there exists an alternative method which is usually preferred for instability problems. This method reduces the instability problems to a characteristic-value problem which, for the present case, is

$$k\frac{d^2\theta}{dx^2} + \beta e^2\theta = 0 \,, \qquad (3.6)$$

$$\theta(0) = \theta(\ell) = 0 \,. \qquad (3.7)$$

This leads to the characteristic functions

$$\theta_n = \sin \lambda_n x \, ,$$ (3.8)

and to the characteristic values

$$\lambda_n \ell = n\pi \, , \quad n = 1, 2, 3, \dots \, ,$$ (3.9)

where

$$\lambda_n = e_n \left(\frac{\beta}{k}\right)^{1/2} = n \left(\frac{\pi}{\ell}\right) \, .$$ (3.10)

The lowest characteristic value,

$$\lambda_1 = e_1 \left(\frac{\beta}{k}\right)^{1/2} = \frac{\pi}{\ell} \, ,$$ (3.11)

gives the electric potential at which the glass temperature becomes unstable.

Note that, in the process of selecting an approximate circular profile for the energy method, we apparently picked the exact solution of the problem. Consequently, both methods resulted in the same answer. This does not usually turn out to be the case for more complicated problems.

So far we have ignored the unsteadiness of the foregoing example. Clearly, in the example taken from mechanics (statics) the dynamic effects are negligible. However, the conduction in a glass plate may be unsteady, which requires the consideration of the unsteady thermal energy,

$$\rho c \frac{\partial \theta}{\partial t} = k \frac{\partial^2 \theta}{\partial x^2} + \beta e^2 \theta \, .$$ (3.12)

We now wish to determine the effect of unsteadiness on the stability of the plate temperature. Let

$$\theta_n = A_n e^{\sigma_n t} \sin \lambda_n x \, ,$$ (3.13)

where λ_n is as given by Eq. (3.10), A_n is the amplitude, and

$$\sigma_n = 0$$ (3.14)

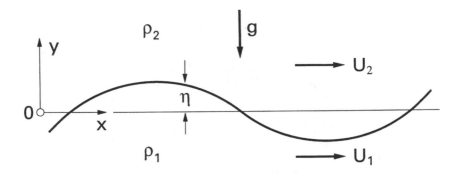

FIGURE 3.4: Interface of two fluids.

determines the (marginal) steady state. Equation (3.13) inserted into Eq. (3.12) yields the same stability condition,

$$\frac{\sigma_n}{\alpha} = \left[e_n^2 \left(\frac{\beta}{k} \right) - n^2 \left(\frac{\pi}{\ell} \right)^2 \right] = 0 \,. \tag{3.15}$$

So far we learned the foundations of stability theory in terms of two illustrative examples. Actually, out interest lies in flow instability which we consider next.

3.1   INSTABILITY

Foregoing elementary examples somewhat conceal the complexities involved with instability problems. The next example on the interface instability of two semi-infinite horizontal fluid layers illustrates this fact.

Consider two incompressible fluids of densities ρ_1, ρ_2, one beneath the other, moving with horizontal velocities U_1, U_2, respectively (Fig. 3.4) Neglect viscosity and surface tension. We wish to study the interface stability.

The problem satisfies the following four conditions: i) $\boldsymbol{f} = -\boldsymbol{\nabla}\Omega$ with $\Omega = gy$, ii) Inviscid flow, iii) Incompressible fluid and iv) Potential flow, $\boldsymbol{V} = -\boldsymbol{\nabla}\psi$. The conservation of mass gives

$$\boldsymbol{\nabla} \cdot \boldsymbol{V} = 0 \,, \quad \boldsymbol{\nabla} \cdot (\boldsymbol{\nabla}\psi) = 0 \,, \quad \nabla^2\psi = 0 \,. \tag{3.16}$$

The momentum balance,

$$\rho \frac{DV}{Dt} = \rho f - \nabla p \, , \tag{3.17}$$

noting

$$\omega = \nabla \times V = \nabla \times \nabla \psi = 0 \, ,$$

$$\frac{DV}{Dt} = \frac{\partial V}{\partial t} + \omega \times V + \nabla \left(\frac{V^2}{2} \right) , \tag{3.18}$$

and $f = -\nabla \Omega$, may be rearranged as

$$\frac{\partial V}{\partial t} + \nabla \left(\frac{V^2}{2} \right) = -\nabla \Omega - \nabla \left(\frac{p}{\rho} \right) , \tag{3.19}$$

or,

$$\frac{\partial V}{\partial t} + \nabla \left(p/\rho + V^2/2 + \Omega \right) = 0 \, . \tag{3.20}$$

Furthermore, since $V = -\nabla \psi$ and $\Omega = gy$,

$$-\nabla \frac{\partial \psi}{\partial t} + \nabla \left(p/\rho + V^2/2 + \Omega \right) = 0 \tag{3.21}$$

or

$$-\frac{\partial \psi}{\partial t} + \left(\frac{p}{\rho} + \frac{V^2}{2} + gy \right) = F(t) \tag{3.22}$$

which is the *second Bernoulli* theorem.

For small disturbances about the initial steady flow,

$$\psi = -Ux + \phi \, (\text{disturbance}) \, , \tag{3.23}$$

and noting $u = -\partial \psi / \partial x$ and $v = -\partial \psi / \partial y$, rearrange Bernoulli as

$$\frac{p}{\rho} = \frac{\partial \phi}{\partial t} - \frac{1}{2} \left[\left(U - \frac{\partial \phi}{\partial x} \right)^2 + \left(\frac{\partial \phi}{\partial y} \right)^2 \right] - gy + F(t) \tag{3.24}$$

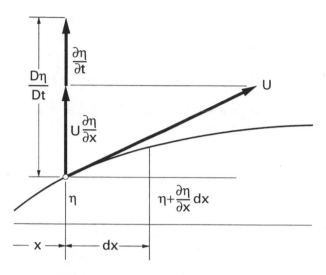

FIGURE 3.5: Velocity continuity across interface.

which yields, after neglecting higher order terms and separating the base flow,

$$\frac{p}{\rho} = \frac{\partial \phi}{\partial t} + U \frac{\partial \phi}{\partial x} - gy + \cdots .$$

(3.25)

Then, from *pressure continuity* across interface,

$$\rho_1 \left(\frac{\partial \phi_1}{\partial t} + U_1 \frac{\partial \phi_1}{\partial x} - g\eta \right) = \rho_2 \left(\frac{\partial \phi_2}{\partial t} + U_2 \frac{\partial \phi_2}{\partial x} - g\eta \right) .$$

(3.26)

For *velocity continuity* across interface, assume the velocity of each fluid at the interface is made up of the velocity of interface, and the velocity of the fluid relative to it. Hence, for the vertical component (Fig. 3.5)

$$\frac{D\eta}{Dt} = \frac{\partial \eta}{\partial t} + U \frac{\partial \eta}{\partial x} = V = -\frac{\partial \phi}{\partial y} ,$$

(3.27)

and, for each fluid,

$$\frac{\partial \eta}{\partial t} + U_1 \frac{\partial \eta}{\partial x} = -\frac{\partial \phi_1}{\partial y} , \quad \frac{\partial \eta}{\partial t} + U_2 \frac{\partial \eta}{\partial x} = -\frac{\partial \phi_2}{\partial y}$$

(3.28)

which are the kinematic (compatibility) conditions to be satisfied for $y = 0$.

We now seek solutions of

$$\nabla^2 \phi = \frac{\partial^2 \phi}{\partial x^2} + \frac{\partial^2 \phi}{\partial y^2} = 0 , \quad \phi = \phi(x, y, t) \tag{3.29}$$

satisfying $\nabla \phi \to 0$ as $y \to \pm\infty$ and oscillating ϕ in x. Selecting the sign of the separation parameter accordingly,

$$\phi(x, y, t) = A(t) e^{k(\pm y - ix)} \tag{3.30}$$

where $A(t)$ denotes the time-dependence of disturbances. Furthermore, the linearized problem suggests an exponential behavior for $A(t)$. In view of this fact, let

$$\phi_1 = C_1 e^{ky} e^{i(\sigma t - kx)} , \quad \phi_2 = C_2 e^{-ky} e^{i(\sigma t - kx)} , \tag{3.31}$$

and

$$\eta = A e^{i(\sigma t - kx)} . \tag{3.32}$$

Inserting Eqs. (3.31) and (3.32) into Eqs. (3.26) and (3.28) we get

$$\rho_1 \left[i(\sigma - kU_1)C_1 - gA \right] = \rho_2 \left[i(\sigma - kU_2)C_2 - gA \right] \tag{3.33}$$

and

$$i(\sigma - kU_1)A = -kC_1 , \quad i(\sigma - kU_2)A = +kC_2 \tag{3.34}$$

Eliminating C_1 and C_2 among Eqs. (3.33) and (3.34), we have

$$\rho_1(\sigma - kU_1)^2 + \rho_2(\sigma - kU_2)^2 = gk(\rho_1 - \rho_2) , \tag{3.35}$$

or, after rearrangement,

$$(\rho_1 + \rho_2)\sigma^2 - 2k(\rho_1 U_1 + \rho_2 U_2)\sigma$$
$$+ k^2(\rho_1 U_1^2 + \rho_2 U_2^2) - gk(\rho_1 - \rho_2) = 0 . \tag{3.36}$$

The roots of Eq. (3.36) are

$$\underbrace{\frac{\sigma}{k}}_{C} = \underbrace{\frac{\rho_1 U_1 + \rho_2 U_2}{\rho_1 + \rho_2}}_{C_M} \pm \underbrace{\left[\underbrace{\frac{g}{k}\left(\frac{\rho_1 - \rho_2}{\rho_1 + \rho_2} \right)}_{C_0^2} - \frac{\rho_1 \rho_2}{(\rho_1 + \rho_2)^2}(U_1 - U_2)^2 \right]^{1/2}}_{C_R} \tag{3.37}$$

C_M being the mean velocity of the two currents. Relative to this velocity, waves travel with velocities $\pm C_R$, and C_0 denotes the wave velocity in the absence of or equal currents. Note that the values of σ are imaginary for negative values of the bracketed terms, and interface is unstable. The case for $U_1 = U_2 = 0$ and $\rho_1 < \rho_2$ is called the *Rayleigh-Taylor* instability, and the case for $\rho_1 = \rho_2$ or $g = 0$ called the *Kelvin-Helmholtz* instability.

It is interesting to note from energy considerations, the following intuitive arguments,

$$\text{Kinetic energy} \sim \text{Potential energy},$$

$$\left(m\frac{V}{t} \sim F \right) V,$$

where

$$m \sim (\rho_1 + \rho_2)\ell^3,$$

$$\sigma \sim \frac{1}{t}, \quad \ell \sim \frac{1}{k}, \quad \frac{V}{t} \sim \frac{\ell}{t^2} \sim \frac{\sigma^2}{k}$$

$$F \sim g(\rho_1 - \rho_2)\ell^3$$

and

$$\frac{\sigma}{k} \sim \left[\frac{g}{k} \left(\frac{\rho_1 - \rho_2}{\rho_1 + \rho_2} \right) \right]^{1/2} \tag{3.38}$$

which is identical to the first ratio in brackets in Eq. (3.37). For the second ratio, consider two semi-infinite fluid layers of equal density, initially having different uniform velocities, U_1 and U_2 (Fig. 3.6). Bring the layers together and let the interface finally assume mean velocity $U_m = (U_1 + U_2)/2$. From the difference between the final and initial kinetic energies,

$$2m \left(\frac{U_m^2}{2} \right) - \frac{m}{2} \left(U_1^2 + U_2^2 \right), \tag{3.39}$$

one readily gets

$$\frac{\sigma}{k} \sim \left[-\frac{(U_1 - U_2)^2}{4} \right]^{1/2} \tag{3.40}$$

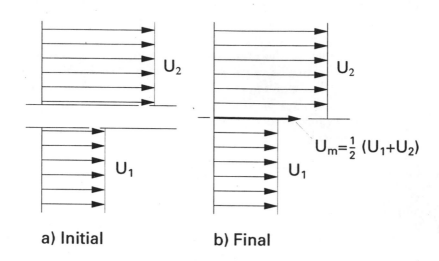

a) Initial   **b) Final**

FIGURE 3.6: (a) Initial condition (b) Final condition.

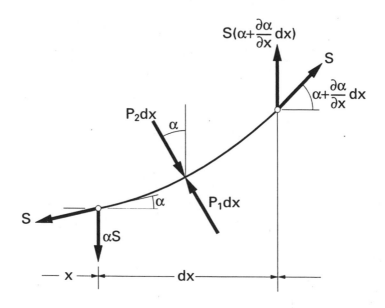

FIGURE 3.7: Effect of surface tension.

which is identical to the second term in brackets for the case of $\rho_1 = \rho_2$. This is a very important result. It shows that *an inviscid fluid layer subject to a velocity gradient is unstable.*

For the effect of surface tension, modify the interface pressure continuity (Fig. 3.7),

$$-S\alpha - p_2 \, dx + p_1 \, dx + S\left(\alpha + \frac{\partial\alpha}{\partial x}dx\right) = 0 \qquad (3.41)$$

which gives

$$p_1 - p_2 + S\frac{\partial\alpha}{\partial x} = 0 , \qquad (3.42)$$

or, in terms of

$$\alpha = \frac{\partial\eta}{\partial x} , \qquad (3.43)$$

$$p_1 - p_2 + S\frac{\partial^2\eta}{\partial x^2} = 0 . \qquad (3.44)$$

Now employing Eq. (3.44) in place of Eq. (3.26), we obtain

$$\rho_1\left[i(\sigma - kU_1)C_1 - gA\right] - k^2 AS = \rho_2\left[i(\sigma - kU_2)C_2 - gA\right] , \qquad (3.45)$$

which leads to

$$(\rho_1 + \rho_2)\sigma^2 - 2k(\rho_1 U_1 + \rho_2 U_2)\sigma$$
$$+ k^2(\rho_1 U_1^2 + \rho_2 U_2^2) - gk(\rho_1 - \rho_2) - k^3 S = 0 \qquad (3.46)$$

which gives

$$\frac{\sigma}{k} = \frac{\rho_1 U_1 + \rho_2 U_2}{\rho_1 + \rho_2}$$
$$\pm \left[\frac{g}{k}\left(\frac{\rho_1 - \rho_2}{\rho_1 + \rho_2}\right) + \frac{kS}{(\rho_1 + \rho_2)} - \frac{\rho_1\rho_2}{(\rho_1 + \rho_2)^2}(U_1 - U_2)^2\right]^{1/2} , \qquad (3.47)$$

where the wave velocity for stagnant fluids is

$$C_0^2 = \frac{g}{k}\left(\frac{\rho_1 - \rho_2}{\rho_1 + \rho_2}\right) + \frac{kS}{(\rho_1 + \rho_2)} . \qquad (3.48)$$

So far we assumed that $\rho_1 > \rho_2$ is necessary for stability when surface tension is neglected. Eq. (3.48) shows that there is stability even when $\rho_1 < \rho_2$, provided

$$\lambda/2\pi < [S/g(\rho_2 - \rho_1)]^{1/2} . \tag{3.49}$$

For the effect of viscosity, we follow an approximate intuitive approach rather than the formal development which is somewhat involved but available in the literature (see, for example, Chandrasekhar 1961 or Drazin and Reid 1981). Consider the case of $U_1 = U_2 = 0$. Under the influence of viscosity, the damped interface is expected to satisfy approximately

$$\frac{d^2 A_k}{dt^2} + f(\nu, k)\frac{dA_k}{dt} + \omega^2 A_k = 0 \tag{3.50}$$

where

$$f(\nu, k) \sim \frac{1}{t} \sim \nu k^2 , \quad \omega \sim \frac{1}{t} \sim (bk)^{1/2} , \tag{3.51}$$

and b is an effective acceleration, say

$$b = g\left(\frac{\rho_1 - \rho_2}{\rho_1 + \rho_2}\right) + \frac{k^2 S}{\rho_1 + \rho_2} \tag{3.52}$$

which can be readily seen from Eq. (3.47). Then, with an unknown coefficient C, Eq. (3.50) becomes

$$\frac{d^2 A_k}{dt^2} + C\nu k^2 \frac{dA_k}{dt} + \omega^2 A_k = 0 . \tag{3.53}$$

To proceed further, we consider the two limits of Eq. (3.53) corresponding to small and large dampings. We know from the literature that, for small damping (long-wave limit),

$$C = 4 , \quad \nu k^2 \ll \omega^2 \tag{3.54}$$

and for large damping (small-wave limit),

$$C = 2 , \quad \nu k^2 \gg \omega^2 \tag{3.55}$$

and the exact solution of the problem shows that Eq. (3.55) continues to hold even for

$$\nu k^2 \sim \omega^2 . \tag{3.56}$$

For an unstable interface ($\omega^2 = -\sigma^2$), the growth of the amplitude is expected to be

$$A_k(t) = A(0) \exp(nt) . \tag{3.57}$$

Then, Eq. (3.53) for short waves is reduced to

$$n^2 + 2\nu k^2 n - bk = 0 \tag{3.58}$$

and

$$\frac{dn}{dk} = 0 \tag{3.59}$$

gives the wavelength corresponding to the most rapid growth of interface

$$\lambda_{\max} = 4\pi \left(\frac{\nu^2}{b}\right)^{1/3} . \tag{3.60}$$

Also, the case of long waves leads to

$$\lambda_{\max} = 4\pi \left(\frac{4\nu^2}{b}\right)^{1/3} . \tag{3.61}$$

In Chapter 4 we relate these wavelengths to a Kolmogorov structure.

So far, as we have seen in the foregoing examples, disturbances in unstable flows determined by the method of small perturbations grow exponentially with time. In reality, however, disturbances quickly grow beyond the exponential range and the theory fails. Besides, any extension of the range of the infinitesimal theory by a finite amplitude theory turns out to be far from describing large fluctuations needed for turbulence (Fig. 3.8). *An instability theory can only be considered as the origin of but not for an answer to turbulence.*

Before looking for the answer, let us momentarily go back to the illustrative example of Chapter 1 (see p. 2) dealing with sudden boundary acceleration of a semi-infinite fluid. Replace the acceleration with a periodic oscillation, say,

$$u = U \cos \omega t .$$

The solution of the problem (see, for example, Ex. 7-28 in Arpacı 1966),

$$\frac{u}{U} = \exp\left[\left(\frac{\omega}{2\nu}\right)^{1/2} y\right] \cos\left[\omega t - \left(\frac{\omega}{2\nu}\right)^{1/2} y\right]$$

$$-\frac{1}{\pi} \int_0^\infty \frac{re^{-tr} \sin(r/\nu)^{1/2} y}{(r^2 + \omega^2)} \, dr ,$$

has a somewhat difficult, unsteady periodic part (involving the integral) that vanishes as $t \to \infty$, and a relatively simple, steady periodic part. Generalizing this fact, we may state that *a developing flow* (laminar or turbulent) *is more involved* (and less useful for all practical purposes) *than its fully developed limit.* This realization suggests that we proceed directly to a study of fully developed turbulent flows and related heat and mass transfer.

3.2   REYNOLDS STRESS

Assume that the instantaneous velocity of turbulence satisfies the usual momentum equations, which are, for an incompressible fluid,

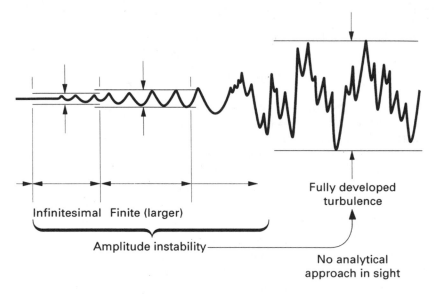

Infinitesimal Finite (larger)

Amplitude instability

Fully developed turbulence

No analytical approach in sight

FIGURE 3.8: Origin of turbulence.

$$\frac{\partial \tilde{u}_i}{\partial t} + \tilde{u}_j \frac{\partial \tilde{u}_i}{\partial x_j} = \frac{1}{\rho}\frac{\partial}{\partial x_j}\tilde{\sigma}_{ij} \,, \qquad (3.62)$$

where $\tilde{\sigma}_{ij}$ is the stress tensor. For an incompressible Newtonian fluid,

$$\tilde{\sigma}_{ij} = -\tilde{p}\delta_{ij} + 2\mu\tilde{s}_{ij} \,, \qquad (3.63)$$

$\tilde{s}_{ij}$ being the rate of strain tensor, defined by

$$\tilde{s}_{ij} = \frac{1}{2}\left(\frac{\partial \tilde{u}_i}{\partial x_j} + \frac{\partial \tilde{u}_j}{\partial x_i}\right) , \qquad (3.64)$$

Inserting Eq. (3.63) into Eq. (3.62) utilizing the conservation of mass, $\partial \tilde{u}_i/\partial x_i = 0$, we have the Navier-Stokes equations for an incompressible fluid,

$$\frac{\partial \tilde{u}_i}{\partial t} + \tilde{u}_j\frac{\partial \tilde{u}_i}{\partial x_j} = -\frac{1}{\rho}\frac{\partial \tilde{p}}{\partial x_j} + \nu\nabla^2\tilde{u}_i . \qquad (3.65)$$

There are two approaches to the integration of Eq. (3.65), the statistical and computational. The present state of the statistical approach, including recent impressive advances, is far from providing an answer to turbulence. The long-time integration of Eq. (3.65) by computational means is also beyond the capacity of presently available as well as projected future computers. If there is no statistical or computational method available for studying instantaneous turbulence, what about its average over a meaningful time interval? Actually, it is not the instantaneous turbulence, for example, in a pipe, but rather its mean value, which provides the net flow through this pipe that has prime technological importance. Instantaneous turbulence, and its mean value relative to laminar flow near the walls of a pipe, are sketched in Fig. 3.9.

What we need now are the Navier-Stokes equations satisfied by the mean turbulence. First, consider the instantaneous velocity $\tilde{u}_i$ expressed as velocity fluctuations u_i superimposed on the mean flow U_i such that

$$\tilde{u}_i = U_i + u_i , \qquad (3.66)$$

where

$$\overline{\tilde{u}_i} = U_i = \lim_{T\to\infty}\frac{1}{T}\int_{t_0}^{t_0+T} \tilde{u}_i \, dt . \qquad (3.67)$$

Hereafter, capital letters are used for mean values and lower-case letters for fluctuating values. An overbar denotes averaging, and we confine our interest to turbulence that is steady on the mean, hence $\partial U_i/\partial t = 0$.

Because of the nonlinearity of the Navier-Stokes equations, we need the average of the product of two instantaneous variables. To develop

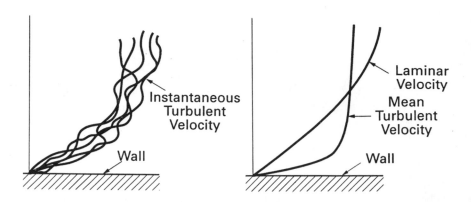

FIGURE 3.9: (a) Instantaneous turbulence (b) Mean turbulence.

this average, consider any two variables (scalar, vector or tensor), say $\tilde{a}_1$ and $\tilde{a}_2$. In terms of means and fluctuations,

$$\tilde{a}_1 = A_1 + a_1 , \quad \tilde{a}_2 = A_2 + a_2 .$$

By definition,

$$\overline{\tilde{a}}_1 = A_1 , \quad \overline{\tilde{a}}_2 = A_2$$

and, consequently,

$$\overline{a}_1 = 0 , \quad \overline{a}_2 = 0 .$$

Then, the average of $\tilde{a}_1 \tilde{a}_2$,

$$\overline{\tilde{a}_1 \tilde{a}_2} = \overline{(A_1 + a_1)(A_2 + a_2)} = A_1 A_2 + A_1 \overline{a}_2 + \overline{a}_1 A_2 + \overline{a_1 a_2}$$

gives

$$\overline{\tilde{a}_1 \tilde{a}_2} = A_1 A_2 + \overline{a_1 a_2} . \tag{3.68}$$

Now, we are ready for the average of the instantaneous Navier-Stokes equations. In addition to $\tilde{u}_i = U_i + u_i$, introducing

$$\tilde{p} = P + p , \quad \overline{p} = 0 ,$$

$$\tilde{\sigma}_{ij} = \Sigma_{ij} + \sigma_{ij} , \quad \overline{\sigma}_{ij} = 0 ,$$

$$\Sigma_{ij} = -P\delta_{ij} + 2\mu S_{ij} \, ,$$

and noting, in view of Eq. (3.68) and $\partial \tilde{u}_j / \partial x_j = 0$, that

$$\overline{\tilde{u}_j \frac{\partial \tilde{u}_i}{\partial x_j}} = \overline{(U_j + u_j)\frac{\partial}{\partial x_j}(U_i + u_i)} = U_j \frac{\partial U_i}{\partial x_j} + \frac{\partial}{\partial x_j}(\overline{u_i u_j}) \, , \quad (3.69)$$

the average of Eq. (3.62) may be written as

$$U_j \frac{\partial U_i}{\partial x_j} = \frac{1}{\rho} \frac{\partial}{\partial x_j} (\Sigma_{ij} - \rho \overline{u_i u_j}) \, , \qquad (3.70)$$

where

$$T_{ij} = \Sigma_{ij} - \rho \overline{u_i u_j} = -P\delta_{ij} + 2\mu S_{ij} - \rho \overline{u_i u_j} \qquad (3.71)$$

is the total mean stress in a turbulent flow. The contribution of the turbulent motion to the mean stress,

$$\boxed{\tau_{ij}^R = -\rho \overline{u_i u_j}} \, , \qquad (3.72)$$

is called the Reynolds stress. This stress is symmetric, $\tau_{ij}^R = \tau_{ji}^R$. The diagonal components of τ_{ij}^R are normal stresses (negative pressures), and off-diagonal components are shear stresses. The latter play an important role in the mean momentum transfer by turbulent motion.

Having averaged the momentum equations, we proceed to a similar averaging of the rate of thermal energy.

3.3   REYNOLDS FLUX

Turbulent flow transports passive contaminants such as chemical species, small particles, and fluid properties such as thermal energy in much the same way as momentum. For use in thermal models, we develop here the equation governing the turbulent transport of thermal energy.

The diffusion of thermal energy in an incompressible turbulent flow satisfies, after neglecting the viscous dissipation,

$$\frac{\partial \tilde{\theta}}{\partial t} + \tilde{u}_i \frac{\partial \tilde{\theta}}{\partial x_i} = \alpha \frac{\partial^2 \tilde{\theta}}{\partial x_i \partial x_i} \, , \qquad (3.73)$$

where α is the thermal diffusivity, which will be assumed constant. Decompose $\tilde{\theta}(x_i, t)$ into a mean value Θ and fluctuation θ as done for $\tilde{u}_i(x_i, t)$,

$$\tilde{\theta} = \Theta + \theta \, , \qquad (3.74)$$

where

$$\bar{\theta} = \Theta = \lim_{T \to \infty} \frac{1}{T} \int_{t_0}^{t_0+T} \tilde{\theta} \, dt \, , \qquad (3.75)$$

$$\bar{\theta} = 0 \, , \quad \frac{\partial \Theta}{\partial t} = 0 \, , \qquad (3.76)$$

the last condition confining our interest to the steady mean turbulence.

Inserting Eq. (3.74) into Eq. (3.73) and taking the average of all terms in the resulting equation yields

$$U_i \frac{\partial \Theta}{\partial x_i} = \frac{\partial}{\partial x_i} \left(\alpha \frac{\partial \Theta}{\partial x_i} - \overline{u_i \theta} \right) \, . \qquad (3.77)$$

The total mean heat flux q_i in a turbulent flow then becomes

$$q_i = \rho c_p \left(-\alpha \frac{\partial \Theta}{\partial x_i} + \overline{u_i \theta} \right) \, . \qquad (3.78)$$

This result shows that the turbulent heat flux is a sum involving contributions from molecular motion and turbulent motion. The contribution of turbulent motion to the mean heat flux

$$\boxed{q_i^R = \rho c_p \overline{u_i \theta}} \qquad (3.79)$$

is called the Reynolds flux. The analogy between Eqs. (3.72) and (3.79) should be noted. This analogy is the foundation of the assumption that turbulence transports thermal energy in the same way as momentum.

Let us summarize in Table 3.1 what we have learned so far. For instantaneous turbulence, the formulation is complete. We have five equations (conservation of mass, three components of the balance of momentum, and thermal energy) and five unknowns ($\tilde{p}, \tilde{u}_i, \tilde{\theta}$). However, there is no solution (by any method) resulting from long-time integration of this formulation. Long-time integration of nonlinear equations is still in its stage of infancy. On the other hand, for the mean (steady) turbulence, the formulation becomes incomplete because we continue to have five equations but now have additional unknowns τ_{ij}^R and q_i^R. However, in terms of models for τ_{ij}^R and q_i^R, this formulation becomes complete and can be solved. We have then a steady problem which does not require any temporal integration.

TABLE 3.1:

Instantaneous turbulence
$$\frac{\partial \tilde{u}_i}{\partial x_i} = 0$$ $$\frac{\partial \tilde{u}_i}{\partial t} + \tilde{u}_j \frac{\partial \tilde{u}_i}{\partial x_j} = -\frac{1}{\rho}\frac{\partial \tilde{p}}{\partial x_i} + \nu \nabla^2 \tilde{u}_i$$ $$\frac{\partial \tilde{\theta}}{\partial t} + \tilde{u}_j \frac{\partial \tilde{\theta}}{\partial x_j} = \alpha \nabla^2 \tilde{\theta}$$
Complete formulation (five equations and five unknowns). No solution (by any method) is in sight.

Mean (steady) turbulence
$$\frac{\partial U_i}{\partial x_i} = 0$$ $$U_j \frac{\partial U_i}{\partial x_j} = -\frac{1}{\rho}\frac{\partial P}{\partial x_i} + \nu \nabla^2 U_i + \frac{1}{\rho}\frac{\partial \tau_{ij}^R}{\partial x_j}$$ $$U_i \frac{\partial \Theta}{\partial x_i} = \alpha \nabla^2 \Theta - \frac{1}{\rho c_p}\frac{\partial q_i^R}{\partial x_i}$$
Incomplete formulation (because of additional unknowns τ_{ij}^R and q_i^R). A solution is possible after models for $\tau_{ij}^R = -\rho \overline{u_u u_j}$ and $q_i^R = \rho c_p \overline{u_i \theta}$.

The appearance of extra unknowns τ_{ij}^R and q_i^R, which result from the averaging of the instantaneous formulation, is the closure problem of turbulence. However, it should be emphasized that, any time or space averaging of a nonlinear (laminar or turbulent) formulation generates additional unknowns which are explicitly recognized and explored in the case of turbulent flows. The closure problem, that is trying to complete the formulation of the mean steady turbulence by assuming models for $\tau_{ij}^R = -\rho\overline{u_i u_j}$ and $q_i^R = \rho c_p \overline{u_i \theta}$ is a rather difficult task related to the *structure of turbulence* which is the concern of most research on turbulence. We shall circumvent this task by studying the *universal microscales of turbulence* which are the concern of the next chapter.

CHAPTER 4

SCALES OF TURBULENT FLOWS

In Chapter 2 we distinguished three methods for dimensional analysis based on

   i) Nondimensionalized Governing Equations,

   ii) Modified Π-theorem,

   iii) Physical Similitude.

Employing only *one-scale* (for length, time, velocity and temperature), we established relations among (dependent and independent) dimensionless numbers such as

$$Nu = f(Re, Pr) , \quad Nu = f(Ra, Pr) , \quad \text{etc.} \qquad (4.1)$$

A dimensional analysis based on one-scale for each variable leads to an implicit relation among dimensionless numbers. In contrast, a dimensional analysis based on multi-scales for at least one variable leads to an explicit relation among dimensionless numbers such as

$$Nu \sim Re^{1/2} Pr^{1/3} , \quad Nu \sim Ra^{1/4} , \quad \text{etc.} \qquad (4.2)$$

Our concern in this chapter is multi-scale dimensional analysis of turbulent flows. Before getting involved with the complexity of these flows, however, let us first illustrate the foundations of multi-scale dimensional analyses in terms of a couple examples involving laminar flows.

Consider first the steady flow with free stream velocity U_∞ and temperature T_∞ of an incompressible fluid over an horizontal plate kept at uniform temperature T_w.

On dimensional grounds, we have for momentum

$$F_I \sim F_V \qquad (4.3)$$

and for thermal energy (rate of)

$$Q_H \sim Q_K \qquad (4.4)$$

in terms of the notation used in Chapter 2. Let us nondimensionalize Eqs. (4.3) and (4.4) by an integral (or geometric) scale (ℓ) and two diffusion scales (δ, δ_θ),

$$U_\infty \frac{U_\infty}{\ell} \sim \nu \frac{U_\infty}{\delta^2} , \qquad (4.5)$$

and

$$U_\theta \frac{\Delta T}{\ell} \sim \alpha \frac{\Delta T}{\delta_\theta^2} , \qquad (4.6)$$

where $\Delta T = T_w - T_\infty$.

For an incompressible fluid, momentum is decoupled from thermal energy and the momentum boundary layer thickness relative to the integral scale obtained from Eq. (4.5) is

$$\frac{\delta}{\ell} \sim \frac{1}{Re^{1/2}} , \qquad Re = \frac{U_\infty \ell}{\nu} . \qquad (4.7)$$

Then, for the friction coefficient, we have from

$$f = \frac{\tau_w}{\frac{1}{2}\rho U_\infty^2} \sim \frac{\mu \dfrac{U_\infty}{\delta}}{\frac{1}{2}\rho U_\infty^2} \sim \frac{\nu}{U_\infty \delta} = \frac{1}{Re}\left(\frac{\ell}{\delta}\right) \qquad (4.8)$$

which yields, in terms of Eq. (4.7),

$$f \sim \frac{1}{Re^{1/2}} . \qquad (4.9)$$

For heat transfer from the plate, the boundary conduction related to convection,

$$k \frac{\Delta T}{\delta_\theta} \sim h \Delta T , \qquad (4.10)$$

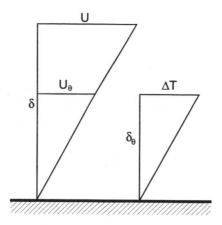

FIGURE 4.1: Analogy between momentum and thermal energy.

h being the heat transfer coefficient, gives

$$Nu = \frac{h\ell}{k} \sim \frac{\ell}{\delta_\theta} \tag{4.11}$$

which can be rearranged as

$$Nu \sim \left(\frac{\ell}{\delta}\right) \frac{\delta}{\delta_\theta} . \tag{4.12}$$

Now, using the analogy between momentum and thermal energy[1], we have from the ratio between Eqs. (4.5) and (4.6),

$$\frac{U_\infty}{U_\theta} = \frac{\nu}{\alpha} \left(\frac{\delta_\theta}{\delta}\right)^2 , \tag{4.13}$$

and from Fig. 4.1,

$$\frac{U_\infty}{U_\theta} = \frac{\delta}{\delta_\theta} . \tag{4.14}$$

(One needs to justify using Eq. 4.14 based on linear profiles in place of actual velocity and temperature which are not linear!). Elimination of U_∞/U_θ between Eqs. (4.13) and (4.14) yields

$$\frac{\delta}{\delta_\theta} = Pr^{1/3} , \tag{4.15}$$

---

[1]Similarity of governing equations and boundary conditions.

where $Pr = \nu/\alpha$ is the Prandtl number. Finally, inserting Eq. (4.15) and the inverse of Eq. (4.7) into Eq. (4.12), we obtain

$$Nu \sim Re^{1/2}Pr^{1/3} \tag{4.16}$$

for the convective heat transfer in forced flow over a horizontal plate.

Next, we proceed to another example illustrating natural convection along a vertical plate kept at temperature T_w above ambient temperature T_∞.

On dimensional grounds, we have for momentum, after neglecting inertial effect[2],

$$F_B \sim F_V \tag{4.17}$$

and for thermal energy (which retains its earlier form)

$$Q_H \sim Q_K . \tag{4.4}$$

In terms of the scales used in the preceding example, except for velocity U which corresponds to $U_{\max}$ here, we have for momentum,

$$g\left(\frac{\Delta\rho}{\rho}\right) \sim \nu\frac{U}{\delta_v^2} , \tag{4.18}$$

and for thermal energy,

$$U\frac{\Delta T}{\ell} \sim \alpha\frac{\Delta T}{\delta_\theta^2} . \tag{4.19}$$

Following Squire (1938), we assume δ_v (not δ in Fig. 4.2) to be important for heat transfer and, furthermore, let

$$\delta_v \sim \delta_\theta \tag{4.20}$$

in Eq. (4.18). Then,

$$U \sim \alpha\frac{\ell}{\delta_\theta^2} \tag{4.21}$$

from Eq. (4.19), inserted into Eq. (4.18), yields

$$g\left(\frac{\Delta\rho}{\rho}\right) \sim \nu\alpha\frac{\ell}{\delta_\theta^4} , \tag{4.22}$$

---

[2]This effect is elaborated following this example.

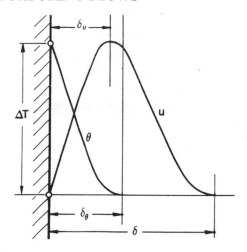

FIGURE 4.2: Momentum, maximum momentum, and thermal boundary layers.

or,

$$Nu \sim \frac{\ell}{\delta_\theta} \sim Ra^{1/4} \tag{4.23}$$

for the convective heat transfer in buoyancy driven flow along a vertical plate. Here

$$Ra = \frac{g}{\nu\alpha} \left(\frac{\Delta\rho}{\rho} \right) \ell^3 \tag{4.24}$$

denotes the Rayleigh number. Using the coefficient of thermal expansion,

$$\beta = \frac{1}{\rho} \left(\frac{\Delta\rho}{\Delta T} \right) , \tag{4.25}$$

this number is more conveniently expressed as

$$Ra = \frac{g\beta\Delta T \ell^3}{\nu\alpha} ,$$

$\Delta T = T_w - T_\infty$ being the specified temperature difference.

Let us see next what happens when we include the inertial effect. Now, momentum becomes

$$F_B \sim F_I + F_V , \tag{4.26}$$

(recall Eq. 2.48), or, on dimensional grounds,

$$g\left(\frac{\Delta\rho}{\rho}\right) \sim u\frac{u}{\ell} + \nu\frac{U}{\delta_\theta^2} \; . \tag{4.27}$$

Then, following the preceding example, and inserting Eq. (4.21) into Eq. (4.27), we get

$$g\left(\frac{\Delta\rho}{\rho}\right) \sim \nu\alpha\frac{\ell}{\delta_\theta^4}\left(1 + \frac{1}{Pr}\right) \; , \tag{4.28}$$

or,

$$Nu \sim \frac{\ell}{\delta_\theta} \sim \Pi_N^{1/4} \; , \tag{4.29}$$

where

$$\Pi_N \sim \frac{Ra}{1 + Pr^{-1}} = \frac{RaPr}{1 + Pr} \tag{4.30}$$

(recall Eq. 2.52). The limit of Eq. (4.29) for $Pr \to \infty$ leads to Eq. (4.23), as expected.

Having developed some appreciation for multi-scale dimensional analysis, we proceed now to turbulent flows. Before the scales associated with these flows, however, we need some ground work which we consider next.

4.1   KINETIC ENERGY OF MEAN FLOW AND OF FLUCTUATIONS

We have already obtained the momentum equations for the steady mean flow of an incompressible fluid (recall Eqs. 3.70 and 3.71),

$$U_j\frac{\partial U_i}{\partial x_j} = \frac{1}{\rho}\frac{\partial T_{ij}}{\partial x_j} \; , \tag{4.31}$$

where

$$T_{ij} = -P\delta_{ij} + 2\mu S_{ij} - \rho\overline{u_iu_j} \; . \tag{4.32}$$

Since the mean momentum $\overline{u}_i$ of turbulent velocity fluctuations is zero, we look for a way other than the mean momentum for a discussion on the interaction between mean flow and turbulence. We shall do this by studying the kinetic energy of mean flow and the mean kinetic energy of turbulence.

The equation governing the mean flow kinetic energy $U_i U_i / 2$ is obtained by multiplying Eq. (4.31) by U_i. After splitting the stress term into two components, we have

$$\rho U_j \frac{\partial}{\partial x_j} \left(\frac{1}{2} U_i U_i \right) = \frac{\partial}{\partial x_j} (T_{ij} U_i) - T_{ij} \frac{\partial U_i}{\partial x_j} . \tag{4.33}$$

Because T_{ij} is symmetric, the last term of Eq. (4.33) may be rearranged as $T_{ij} S_{ij}$,

$$S_{ij} = \frac{1}{2} \left(\frac{\partial U_i}{\partial x_j} + \frac{\partial U_j}{\partial x_i} \right)$$

being the symmetric part of $\partial U_i / \partial x_j$. Then, Eq. (4.33) becomes

$$\rho U_j \frac{\partial}{\partial x_j} \left(\frac{1}{2} U_i U_i \right) = \frac{\partial}{\partial x_j} (T_{ij} U_i) - T_{ij} S_{ij} . \tag{4.34}$$

We have already explored the physical significance of the right-hand terms. Recall from Section 1.3 that dissipation $T_{ij} S_{ij}$ is the deformation part of the work and is the irreversible conversion of kinetic (or mechanical) energy into internal (thermal) energy. Consequently, it is a term in the balance of thermal energy. Reasons for rearranging kinetic energy will become clear in the development of and the discussion following Eq. (4.46).

Inserting Eq. (4.32) into $T_{ij} S_{ij}$ yields

$$T_{ij} S_{ij} = 2\mu S_{ij} S_{ij} - \rho \overline{u_i u_j} S_{ij} . \tag{4.35}$$

Note that the contribution of pressure to deformation work in an incompressible fluid is zero,

$$-P\delta_{ij} S_{ij} = -P S_{ij} = -P \frac{\partial U_i}{\partial x_i} = 0 ,$$

that $2\mu S_{ij} S_{ij}$ is the viscous part and $-\rho \overline{u_i u_j} S_{ij}$ the turbulent part of dissipation in Eq. (4.35). Since turbulent stresses provide the latter, the kinetic energy of turbulence gains from this dissipation. Accordingly, $-\rho \overline{u_i u_j} S_{ij}$ is known as the *turbulent energy production*.

Inserting Eq. (4.32) into Eq. (4.34) yields, after some rearrangement,

$$U_j \frac{\partial}{\partial x_j} \left(\frac{1}{2} U_i U_i \right) = \frac{\partial}{\partial x_j} \left(-\frac{P}{\rho} U_j + 2\nu U_i S_{ij} - \overline{u_i u_j} U_i \right)$$
$$+ \overline{u_i u_j} S_{ij} - 2\nu S_{ij} S_{ij} , \qquad (4.36)$$

the first three terms on the right-hand side being the net work flux associated with mean pressure, mean viscous stress, and turbulent stress, and the last two terms being the dissipation related to turbulent stress and viscous stress, respectively.

Let us introduce now, somewhat loosely, a length scale ℓ, a velocity scale u, and a time scale $t = \ell/u$. Here, u may be defined, for example, as

$$u^2 = \frac{1}{3} \overline{u_i u_i} , \qquad (4.37)$$

and ℓ as an integral scale proportional to a length characterizing geometry. Turbulence scales are indispensable in modeling turbulence. We return to this subject in Section 4.2 and introduce additional scales. In the meantime, the foregoing scales lead to following relations

$$\frac{\partial U_i}{\partial x_i} \sim \frac{u}{\ell} , \quad -\overline{u_i u_j} \sim u^2 , \quad S_{ij} \sim \frac{u}{\ell} ,$$

and

$$\overline{u_i u_j} S_{ij} \sim u\ell S_{ij} S_{ij} , \qquad (4.38)$$

$$-\overline{u_i u_j} U_i \sim u\ell U_i S_{ij} , \qquad (4.39)$$

in which all (mean as well as fluctuating) components are scaled relative to the velocity introduced by Eq. (4.37). We return to and elaborate this assumption in Section 4.4. Now, comparing Eqs. (4.38) and (4.39), respectively, with the first and second viscous terms on the right-hand side of Eq. (4.36), we have

$$\frac{-\overline{u_i u_j} U_i}{2\nu S_{ij} U_i} \sim Re_\ell$$

and

$$\frac{\overline{u_i u_j} S_{ij}}{2\nu S_{ij} S_{ij}} \sim Re_\ell , \qquad (4.40)$$

where $Re_\ell = u\ell/\nu$ is the Reynolds number based on the integral scale. This result shows that the terms associated with turbulent stress are Re_ℓ times larger than those associated with viscous terms. Since this Reynolds number is usually very large, the viscous terms in Eq. (4.36) can ordinarily be neglected. That is, the *gross structure of turbulent flows tends to be independent of viscosity.* This is an important result. However, no further information is apparently available from Eq. (4.36).

We turn now to the mean kinetic energy of turbulent fluctuations. Equation governing this kinetic energy is obtained multiplying Eq. (3.65) by $\tilde{u}_i$, taking the time average of all terms, and subtracting Eq. (4.36). The details of this exercise are left to the reader (see Problem 4-2). The result is

$$U_j \frac{\partial}{\partial x_j}\left(\frac{1}{2}\overline{u_i u_i}\right) = \frac{\partial}{\partial x_j}\left(-\frac{\overline{p}}{\rho}u_j + 2\nu\overline{u_i s_{ij}} - \frac{1}{2}\overline{u_i u_i u_j}\right)$$

$$- \overline{u_i u_j}S_{ij} - 2\nu\overline{s_{ij}s_{ij}} \qquad (4.41)$$

with

$$s_{ij} = \frac{1}{2}\left(\frac{\partial u_i}{\partial x_j} + \frac{\partial u_j}{\partial x_i}\right)$$

for the fluctuating rate of strain. The first three right-hand terms in Eq. (4.41) denote the net work flux associated with fluctuating pressure and fluctuating viscous and turbulent stresses, and the last two terms denote the two kinds of dissipation. Note that turbulent production $-\overline{u_i u_j}S_{ij}$ in Eqs. (4.41) and (4.36) has opposite signs. As expected, this term apparently serves to exchange kinetic energy between mean flow and turbulence. In general, the energy exchange involves a loss to mean flow and a gain to turbulence. The last term in Eq. (4.41) is the rate at which fluctuating viscous stresses perform deformation work against fluctuating strain rate. It denotes a loss of energy, and being quadratic in s_{ij}, is the *viscous dissipation of turbulent fluctuations.* Unlike the dissipation related to mean viscous stress in Eq. (4.36), this term is essential to dynamics of turbulence and cannot be neglected. In view of the foregoing discussion, introducing

$$E = \frac{1}{2}\overline{u_i u_i}\,, \qquad (4.42)$$

for *mean kinetic energy of fluctuations,*

$$\mathcal{D}_j = \frac{\overline{p}}{\rho}u_j - 2\nu\overline{u_i s_{ij}} + \frac{1}{2}\overline{u_i u_i u_j}\,, \qquad (4.43)$$

for *turbulent diffusion* (total work flux associated with fluctuations)

$$\mathcal{P} = -\overline{u_i u_j} S_{ij} \, , \tag{4.44}$$

for *turbulent production*, and

$$\epsilon = 2\nu \overline{s_{ij} s_{ij}} \, , \tag{4.45}$$

for *turbulent dissipation*, Eq. (4.41) may be given the following compact form

$$U_j \frac{\partial E}{\partial x_j} = -\frac{\partial \mathcal{D}_j}{\partial x_j} + \mathcal{P} - \epsilon \tag{4.46}$$

which states that the mean of turbulent kinetic energy fluctuations is balanced by the diffusion, production, and dissipation of this energy.

In a steady, homogeneous, pure shear flow (all averaged quantities except U_i are independent of position and S_{ij} is a constant), Eq. (4.46) reduces to

$$\mathcal{P} = \epsilon \, , \tag{4.47}$$

the balance between the production rate of turbulent energy by Reynolds stresses and the rate of viscous dissipation of turbulent fluctuations. Employing

$$S_{ij} \sim \frac{u}{\ell} \, , \quad -\overline{u_i u_j} \sim u^2$$

and

$$-\overline{u_i u_j} S_{ij} \sim u\ell S_{ij} S_{ij} \, ,$$

this time in Eq. (4.47), we have

$$u\ell S_{ij} S_{ij} \sim \nu \overline{s_{ij} s_{ij}} \, ,$$

or

$$\frac{\overline{s_{ij} s_{ij}}}{S_{ij} S_{ij}} \sim \frac{u\ell}{\nu} = Re_\ell \, . \tag{4.48}$$

Then, in view of the fact that Re, is usually very large, and

$$\overline{s_{ij} s_{ij}} \gg S_{ij} S_{ij} \tag{4.49}$$

which shows that fluctuating strain rate s_{ij} is much larger than mean strain rate S_{ij}. Since strain rates have the dimension of $(\text{time})^{-1}$, the eddies contributing most to the dissipation of energy have very small convection time scales compared to the time scale of the mean flow. Accordingly, the direct interaction between the fluctuating strain rate and the mean strain rate is negligible for large Reynolds numbers.

Foregoing considerations suggest that the scales for fluctuations should be different from those for the mean flow. Accordingly, we may construct a turbulence model to be based on two sets of scales, one set for the mean flow and one set for fluctuations.

4.2 KINETIC SCALES

Let the scales for the mean flow be ℓ, u, and $t = \ell/u$. These scales, being associated with gross behavior of the flow, are the larger ones of the two sets. They apply to the production part of Eq. (4.47). Let the scales for fluctuations be η, v, and $\tau = \eta/v$. These are the smaller ones of the two sets. They apply to the dissipation part of Eq. (4.47). Following Kolmogorov, they are called the (Kolmogorov) microscales. In terms of the integral scales and the microscales, Eq. (4.47) may be written as

$$\mathcal{P} \sim u^2 \frac{u}{\ell} \sim \nu \frac{v^2}{\eta^2} \sim \epsilon \,, \tag{4.50}$$

which illustrates that *viscous dissipation of energy can be estimated from the large-scale inviscid dynamics*. The dissipation may then be interpreted as a passive process proceeding at a rate dictated by the inviscid inertial behavior of large eddies.

For the isotropic limit of the homogeneous flow, we have from Eq. (4.50)

$$u^2 \frac{u}{\ell} \sim \epsilon \,,$$

after letting $\ell \to \eta$ and $u \to v$,

$$v \sim (\eta\epsilon)^{1/3} \,, \tag{4.51}$$

and from the last proportionality of the same equation,

$$v \sim \eta \left(\frac{\epsilon}{\nu}\right)^{1/2} \,. \tag{4.52}$$

Elimination of v between Eqs. (4.51) and (4.52) yields the *Kolmogorov length scale*,

$$\boxed{\eta \sim \left(\frac{\nu^3}{\epsilon}\right)^{1/4}}\,, \tag{4.53}$$

and the insertion of this result into Eq. (4.51) or Eq. (4.52) gives the *Kolmogorov velocity scale*,

$$v \sim (\nu\epsilon)^{1/4}\,. \tag{4.54}$$

Then, for $\tau = \eta/v$, the ratio between Eq. (4.53) and Eq. (4.54) yields the *Kolmogorov time scale*,

$$\tau \sim \left(\frac{\nu}{\epsilon}\right)^{1/2}\,. \tag{4.55}$$

Clearly, Eqs. (4.53)-(4.55) show that small scale motion of turbulence depends only on the momentum diffusivity (kinematic viscosity) and the large-scale energy supply for dissipation ϵ. The Reynolds number in this scale,

$$\boxed{Re_K \sim \frac{v\eta}{\nu} \sim 1}\,, \tag{4.56}$$

illustrates that the small-scale motion is slow and viscous. Foregoing dimensional arguments leading to a two-scale turbulence model are summarized in Table 4.1 for a ready reference.

Actually, a third scale is hidden in the present model. What would happen to η if v were amplified to u of large scales? In other words, what is the length scale associated with

$$\tau^{-1} = \frac{v}{\eta} = \frac{u}{?}\,. \tag{4.57}$$

At first, η is expected to amplify to length ℓ of the large scales. However, from the combination of Eqs. (4.50) and (4.57),

$$u^2\frac{u}{\ell} \sim \nu\frac{v^2}{\eta^2} \sim \nu\frac{u^2}{?^2}\,, \tag{4.58}$$

and from the first and last terms of this result, we have

$$\frac{?}{\ell} \sim \frac{1}{Re_\ell^{1/2}} \ll 1\,, \tag{4.59}$$

TABLE 4.1: A two-scale turbulence model.

Turbulence $\Big\langle$ Mean (large scales) Fluctuations (small scales)
Large scales: $\ell, u, t = \ell/u$, **Small scales:** $\eta, v, \tau = \eta/v$.
Large scale $\equiv$ production scale $\equiv$ integral scale, **Small scale** $\equiv$ dissipation scale $\equiv$ Kolmogorov microscale
Steady homogeneous pure shear flow, $$\mathcal{P} \sim u^2 \frac{u}{\ell} \sim \nu \frac{v^2}{\eta^2} \sim \epsilon ,$$ **provides an inviscid estimate for dissipation.**
Small scales, $$\eta \sim \left(\frac{\nu^3}{\epsilon}\right)^{1/4} , \quad v \sim (\nu\epsilon)^{1/4} , \quad \text{and} \quad \tau \sim \left(\frac{\nu}{\epsilon}\right)^{1/2} ,$$ **are isotropic.**

which shows that the unknown scale must be smaller than the integral scale. Also, according to Eq. (4.57), it is greater than the Kolmogorov scale. Consequently,

$$\eta <? < \ell , \tag{4.60}$$

and the unknown scale turns out to be the *Taylor microscale*, λ. A simple geometric interpretation of Eq. (4.57), now written including the definition of λ as

$$\tau^{-1} = \frac{v}{\eta} = \frac{u}{\lambda} , \tag{4.61}$$

readily follows from the similarity of two shaded triangles shown in Fig. 4.3. Yet a word of caution is in order on this scale. Being defined by the velocity of large scales and the time of small scales, the Taylor scale is

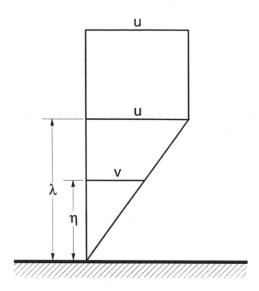

FIGURE 4.3: Geometric interpretation of the Taylor scale.

neither large nor small but rather is an intermediate scale. Frequent use
of this scale in estimates such as $s_{ij} \sim u/\lambda$ is for reasons of convenience
because it involves the measurable velocity of large scales rather than
that of small scales.

Note that Eq. (4.50) may be rearranged in terms of Eq. (4.61) to
give

$$u^2 \frac{u}{\ell} \sim \nu \frac{u^2}{\lambda^2} \,, \tag{4.62}$$

which leads to

$$\frac{\lambda}{\ell} \sim \frac{1}{Re_\lambda} \sim \frac{1}{Re_\ell^{1/2}} \,, \tag{4.63}$$

where $Re_\lambda = u\lambda/\nu$ is the Reynolds number based on the Taylor scale.
Equation (4.50) may also be rearranged in terms of Eq. (4.56) to give

$$\frac{\eta}{\ell} \sim \frac{1}{Re_\eta^3} \sim \frac{1}{Re_\ell^{3/4}} \,. \tag{4.64}$$

Furthermore, from the ratio of Eqs. (4.63) and (4.64), after utilizing the dependence of these equations on Re_ℓ,

$$\frac{\eta}{\lambda} \sim \frac{1}{Re_\eta} \sim \frac{1}{Re_\lambda^{1/2}} \sim \frac{1}{Re_\ell^{1/4}} \, , \qquad (4.65)$$

or,

$$\boxed{Re_\eta \sim Re_\lambda^{1/2} \sim Re_\ell^{1/4}} \, , \qquad (4.66)$$

where $Re = u\eta/\nu$ is the Reynolds number based on the Kolmogorov scale. Finally, an inspection of Eqs. (4.63) and (4.64) also yields the important relation

$$\left(\frac{\eta}{\lambda}\right)^2 \sim \frac{\lambda}{\ell} \qquad (4.67)$$

which helps to construct a turbulence model in the next section.

4.3   A TURBULENCE MODEL

Following earlier developments on electromagnetic waves and kinetic theory of matter, Planck's discovery of photon and de Broglie's discovery of matter waves have completed the theories resting on the particle-wave duality for both radiation and matter. In a similar manner, turbulence research employs an eddy-wave duality in attempts for an understanding of turbulence. Although nothing is wrong with an approach loosely involving such a duality, the eddy of turbulence remained for some years as a large enough blob without any character. However, beginning with earlier works of Thomas and Townsend (1957) and Corrsin (1962), and followed by rapidly growing computational studies, we recognize one of the outstanding features of the small-scale structure of turbulence: *viscous dissipation is not uniformly distributed in space.* This spottiness of dissipation is called the *intermittency* of turbulence. Tennekes (1968) suggests *a simple model for the small-scale structure based on vortex tubes of diameter η stretched by eddies of size λ* (Fig. 4.4). The model assumes the *volume-averaged* dissipation rate to be

$$\epsilon = \nu\frac{u^2}{\lambda^2} \, , \qquad (4.68)$$

the vortex tubes to occupy a fraction,

$$\left(\frac{\eta}{\lambda}\right)^2 \, , \qquad (4.69)$$

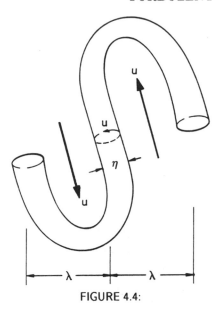

FIGURE 4.4:

of eddy volume λ^3, and the *local* dissipation rate in vortex tubes to be

$$\epsilon = \nu \frac{u^2}{\lambda^2} \left(\frac{\lambda}{\eta}\right)^2 = \nu \frac{u^2}{\eta^2} \,. \qquad (4.70)$$

Also, the model assumes the *local* kinetic energy of vortex tubes to be u^2 (which related to the vortex motion around the axis of a tube) and the *volume-averaged* kinetic energy in vortex tubes to be

$$u^2 \left(\frac{\eta}{\lambda}\right)^2 = u^2 \left(\frac{v}{u}\right)^2 = v^2 \,. \qquad (4.71)$$

In the following section, we condense the foregoing considerations to a simple approach and use it for the rest of the text when looking for scales of complex turbulent flows.

4.4   AN APPROACH TO SCALES

Reconsider the mean kinetic energy of homogeneous velocity fluctuations,

$$\mathcal{P} = \epsilon \qquad (4.72)$$

or, explicitly,

$$-\overline{u_i u_j} S_{ij} = 2\nu \overline{s_{ij} s_{ij}} \ . \tag{4.73}$$

Introduce the following scale relations,

$$u \sim U \ , \tag{4.74}$$

$$-\overline{u_i u_j} \sim u^2 \ , \qquad S_{ij} \sim \frac{u}{\ell} \ , \tag{4.75}$$

$$\overline{s_{ij} s_{ij}} \sim \frac{u^2}{\lambda^2} \ . \tag{4.76}$$

Note that Eq. (4.74) results from

$$u = CU \ , \qquad C = f(\lambda, \ell, U, \nu) \ ,$$

where C depends on *structure* and is not *universal*. As a specific example, recall the Blasius similarity solution of laminar flow over a flat plate which leads to

$$\frac{u}{U} = f'(\eta) \ , \qquad \eta = y \left(\frac{\nu x}{U} \right)^{1/2}$$

where f' satisfies the well-known nonlinear equation. Also the production, being in large scales, is *space-filling* and need to be balanced by the *volume-averaged* dissipation scaled with Eq. (4.76).

In terms of the foregoing relations, Eq. (4.73) becomes

$$\mathcal{P} \sim \frac{u^3}{\ell} \sim \nu \frac{u^2}{\lambda^2} \sim \epsilon \ . \tag{4.77}$$

Equation (4.77) gives, depending on the integral scale,

$$u \sim (\epsilon \ell)^{1/3} \ , \tag{4.78}$$

and, depending on the Taylor scale,

$$u \sim \lambda \left(\frac{\epsilon}{\nu}\right)^{1/2} . \qquad (4.79)$$

Elimination of velocity between Eqs. (4.78) and (4.79) yields the Taylor scale in terms of the integral scale,

$$\lambda \sim \ell^{1/3} \left(\frac{\nu^3}{\epsilon}\right)^{1/6} . \qquad (4.80)$$

Now, consider the *local* (isotropic) behavior of the homogeneous flow (in a sublayer along a wall or in a vortex tube) and let λ, ℓ (which are no longer distinguished) be replaced by an asymptotic scale,

$$\left(\begin{array}{c} \lambda \\ \ell \end{array}\right) \to \eta . \qquad (4.81)$$

Thus, Eq. (4.80) is reduced to

$$\eta \sim \left(\frac{\nu^3}{\epsilon}\right)^{1/4} , \qquad (4.82)$$

Eq. (4.78) or (4.79), in terms of this scale, to

$$v \sim (\nu\epsilon)^{1/4} , \qquad (4.83)$$

and their ratio to

$$\tau \sim \left(\frac{\nu}{\epsilon}\right)^{1/2} . \qquad (4.84)$$

Hereafter, we shall use the foregoing procedure in the construction of microscales. Next, we consider thermal scales associated with turbulent flows.

4.5   THERMAL MEAN AND FLUCTUATIONS

In the preceding sections, we considered (Velocity × Rate of momentum) as a positive measure for mean velocity and turbulent fluctuations. Here, in a similar manner, we consider (Temperature × Rate of internal energy) as a positive measure for thermal mean and fluctuations. Also, note that

$$T\frac{Du}{Dt} = T^2\frac{DS}{Dt}$$

which, in view of $u \sim c_v T$ for an incompressible fluid, is related to the rate of entropy,

$$\frac{c_v}{T^2}\frac{D}{Dt}\left(\frac{1}{2}T^2\right) = \frac{DS}{Dt} .$$

In Section 4.1, the only information we could extract from the mean flow kinetic energy was the *independence of gross kinetic behavior from viscosity*. Here, in a similar manner, it can be shown the *independence of gross thermal behavior from conductivity*. Thus, we proceed directly to thermal fluctuations in a turbulent flow.

4.5.1 Forced Flow

The equation governing $\overline{\theta^2}$ for a steady forced flow is obtained in exactly the same way as the equation for $\overline{u_i u_i}$ (see Problem 4-3). The result is

$$U_i\frac{\partial}{\partial x_j}\left(\frac{1}{2}\overline{\theta^2}\right) = \frac{\partial}{\partial x_i}\left[\frac{1}{2}\overline{u_i\theta^2} - \alpha\frac{\partial}{\partial x_i}\left(\frac{1}{2}\overline{\theta^2}\right)\right]$$
$$- \overline{u_i\theta}\frac{\partial\Theta}{\partial x_i} - \alpha\overline{\frac{\partial\theta}{\partial x_i}\frac{\partial\theta}{\partial x_i}} , \qquad (4.85)$$

or, more compactly,

$$U_i\frac{\partial E_\theta}{\partial x_i} = -\frac{\partial \mathcal{D}_{\theta,i}}{\partial x_i} + \mathcal{P}_\theta - \epsilon_\theta . \qquad (4.86)$$

The flow of $\overline{\theta^2}/2$ is thus controlled by the three right-hand terms, associated with diffusion, production, and dissipation, respectively. Here

$$\mathcal{D}_{\theta,i} = \frac{1}{2}\overline{u_i\theta^2} - \alpha\frac{\partial}{\partial x_i}\left(\frac{1}{2}\overline{\theta^2}\right) \qquad (4.87)$$

may be interpreted as the net flux involving turbulent and molecular contributions. Production and dissipation terms are analogous to those of the turbulent kinetic energy.

For thermal fluctuations of a homogeneous flow (all averaged quantities except Θ are independent of position and $\partial\Theta/\partial x_i$ is a constant)

Eq. (4.85) reduces to

$$-\overline{u_i\theta}\frac{\partial\Theta}{\partial x_i} = \alpha\overline{\frac{\partial\theta}{\partial x_i}\frac{\partial\theta}{\partial x_i}} \, , \tag{4.88}$$

or

$$\mathcal{P}_\theta = \epsilon_\theta \tag{4.89}$$

which states that the production and dissipation of E_θ are balanced. Table 4.3 demonstrates the thermodynamic significance of Eq. (4.88)

Now, letting $\overline{u_i\theta} \sim u_\theta\theta$, $\Theta \sim \theta$ and $\partial\Theta/\partial x_i \sim \theta/\ell$ (where θ may be assumed as the root mean square (rms) of temperature fluctuations), the production term may be rearranged as

$$\mathcal{P}_\theta \sim u_\theta\ell\frac{\partial\Theta}{\partial x_i}\frac{\partial\Theta}{\partial x_i} \, . \tag{4.97}$$

Combining this result with Eq. (4.88) and noting that the Péclet number $Pe_\ell = u_\theta\ell/\alpha$ is usually very large, we obtain

$$\frac{\overline{\dfrac{\partial\theta}{\partial x_i}\dfrac{\partial\theta}{\partial x_i}}}{\dfrac{\partial\Theta}{\partial x_i}\dfrac{\partial\Theta}{\partial x_i}} \sim \frac{u_\theta\ell}{\alpha} = Pe_\ell \gg 1 \, , \tag{4.98}$$

or

$$\overline{\frac{\partial\theta}{\partial x_i}\frac{\partial\theta}{\partial x_i}} \gg \frac{\partial\Theta}{\partial x_i}\frac{\partial\Theta}{\partial x_i} \, . \tag{4.99}$$

That is, the fluctuating temperature gradient is on the average very much larger than the gradient of the mean temperature. This shows, in a manner identical to the discussion following Eq. (4.49), that there is little direct interaction between the large-scale production and the small-scale dissipation of E_θ. Actually, the velocities involved with Eq. (4.88) depend on the Prandtl number and Eq. (4.98) should be corrected accordingly. However, the conclusion reached by Eq. (4.99) continues to remain valid after the correction. We proceed now to the effect of Prandtl number on Eq. (4.88).

Introducing (ℓ, u_θ, θ) for large thermal scales and $(\eta_\theta, v_\theta, \vartheta)$ for small thermal scales, we may express Eq. (4.88) as

$$\mathcal{P}_\theta \sim u_\theta\theta\frac{\theta}{\ell} \sim \alpha\frac{\vartheta^2}{\eta_\theta^2} \sim \epsilon_\theta \, , \tag{4.100}$$

TABLE 4.2: Thermodynamics of Eq. (4.88).

Neglecting work, consider the First Law,

$$dU = \delta Q \,, \qquad (4.90)$$

and express heat flow in terms of entropy flow Ψ,

$$Q = \Psi T \,, \qquad (4.91)$$

so that

$$dU = T\delta\Psi + \Psi\delta T \,. \qquad (4.92)$$

For a positive thermal measure, multiply each term of Eq. (4.92) by T,

$$T\,dU = T^2\delta\Psi + Q\delta T \,. \qquad (4.93)$$

By the Conduction Law,

$$Q \sim -\delta T \,, \qquad (4.94)$$

rearrange Eq. (4.93) as

$$T(dU - T\delta\Psi) \sim -(\delta T)^2 \qquad (4.95)$$

or, in terms of Eqs. (4.91) and (4.92), as

$$-Q\delta T \sim (\delta T)^2 \,. \qquad (4.96)$$

The rate of Eq. (4.96), interpreted by large scales for the left-hand side and by small scales for the right-hand side, provides a thermodynamic interpretation for Eq. (4.88).

where η_θ denotes the thermal counterpart of the Kolmogorov scale. Also, in analogy to Eq. (4.61), introducing a thermal counterpart for the Taylor scale,

$$\frac{\vartheta}{\eta_\theta} = \frac{\theta}{\lambda_\theta} \tag{4.101}$$

Eq. (4.100) may be rearranged as

$$u_\theta \theta \frac{\theta}{\ell} \sim \alpha \frac{\theta^2}{\lambda_\theta^2} . \tag{4.102}$$

To proceed further, we need a velocity scale which depends on Prandtl number. This dependence for fluids with $Pr \geq 1$ (air, water, viscous oils) is different than that for fluids with $Pr \ll 1$ (liquid metals).

First, consider the *case of $Pr \geq 1$*:

From Eq. (4.102), we have

$$u_\theta \sim \alpha \frac{\ell}{\lambda_\theta^2} , \tag{4.103}$$

also, from the similarity between velocity and temperature profiles (recall Fig. 4.1),

$$\frac{u}{\lambda} = \frac{u_\theta}{\lambda_\theta} , \tag{4.104}$$

or, in terms of Eq. (4.103),

$$\frac{u_\theta}{\lambda_\theta} = \frac{u}{\lambda} = \alpha \frac{\ell}{\lambda_\theta^3} . \tag{4.105}$$

Also from Eq. (4.77), a viscous estimate for u,

$$u \sim \lambda \left(\frac{\epsilon}{\nu}\right)^{1/2} . \tag{4.106}$$

Elimination of u/λ between Eqs. (4.105) and (4.106) yields, after some rearrangement,

$$\lambda_\theta \sim \ell^{1/3} \left(\frac{\nu \alpha^2}{\epsilon}\right)^{1/6} , \tag{4.107}$$

and the ratio between Eqs. (4.64) and (4.107) gives

$$\frac{\lambda}{\lambda_\theta} = Pr^{1/3} \ . \tag{4.108}$$

Note that assumed similarity makes Eq. (4.108) an equality. The isotropic limit of Eq. (4.107) for

$$\left(\frac{\lambda_\theta}{\ell} \right) \to \eta_\theta^B \tag{4.109}$$

leads to the *Batchelor scale*,

$$\eta_\theta^B \sim \left(\frac{\nu \alpha^2}{\epsilon} \right)^{1/4} \ . \tag{4.110}$$

The ratio between Eqs. (4.80) and (4.110) gives

$$\frac{\eta}{\eta_\theta^B} = Pr^{1/2} \ . \tag{4.111}$$

As a *second model* for the case of $Pr \geq 1$, consider another limit of the homogeneous flow for

$$\lambda_\theta \to \eta_\theta , \quad \ell \to \eta \ . \tag{4.112}$$

Eq. (4.107) for this limit,

$$\eta_\theta \sim \eta^{1/3} \left(\frac{\nu \alpha^2}{\epsilon} \right)^{1/6} , \tag{4.113}$$

rearranged in terms of η_θ^B, yields

$$\eta_\theta \sim \left[\eta \left(\eta_\theta^B \right)^2 \right]^{1/3} \tag{4.114}$$

which is a *mesomicroscale* based on the weighted geometric average of η and η_θ^B. For any $Pr \geq 1$,

$$\eta_\theta^B \leq \eta_\theta \leq \eta , \tag{4.115}$$

equality holding for the case of $Pr = 1$. The ratio between η and η_θ,

$$\boxed{\frac{\eta}{\eta_\theta} \sim \left(\frac{\eta}{\eta_\theta^B} \right)^{2/3} = Pr^{1/3}} , \tag{4.116}$$

plays a vital role in the following chapters of this text when we interpret the well-known heat and mass transfer correlations by turbulent microscales.

Now, rearrange Eq. (4.67) as

$$\frac{\eta}{\ell} \sim \left(\frac{\lambda}{\ell}\right)^{3/2} , \qquad (4.117)$$

and the ratio between Eq. (4.113) and (4.107) as

$$\frac{\eta_\theta}{\lambda_\theta} \sim \left(\frac{\eta}{\ell}\right)^{1/3} . \qquad (4.118)$$

Then, elimination of η/ℓ between Eqs. (4.117) and (4.118) gives

$$\left(\frac{\eta_\theta}{\lambda_\theta}\right)^2 \sim \frac{\lambda}{\ell} \qquad (4.119)$$

or, in view of Eq. (4.108),

$$\left(\frac{\eta_\theta}{\lambda_\theta}\right)^2 \sim \left(\frac{\lambda_\theta}{\ell}\right) Pr^{1/3} , \quad Pr \geq 1 \qquad (4.120)$$

which helps us to construct a model for *thermal intermittency*. Following Tennekes model for kinetic intermittency, we conjecture conductive tubes of diameter η_θ within vortex tubes of diameter η stretched by thermal eddies of size λ_θ within kinetic eddies of size λ (Fig. 4.5).

Next, consider the *case of $Pr \ll 1$*:
From Eq. (4.50), an inertial estimate for velocity is

$$u \sim (\epsilon \ell)^{1/3} . \qquad (4.121)$$

Using this velocity for that of enthalpy flow, and eliminating u between Eqs. (4.103) and (4.121), we get

$$\lambda_\theta \sim \ell^{1/3} \left(\frac{\alpha^3}{\epsilon}\right)^{1/6} , \qquad (4.122)$$

and from its isotropic limit,

$$\left(\frac{\lambda_\theta}{\ell}\right) \to \eta_\theta^C \qquad (4.123)$$

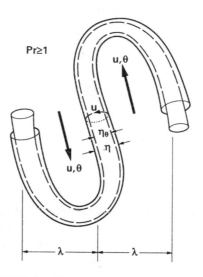

FIGURE 4.5: Thermal intermittency ($Pr \geq 1$).

which leads to the *Oboukhov-Corrsin* scale,

$$\eta_\theta^C \sim \left(\frac{\alpha^3}{\epsilon}\right)^{1/4}. \tag{4.124}$$

Now, it can readily be shown, in a manner similar to ratios among η, λ and ℓ, that

$$\frac{\lambda_\theta}{\ell} \sim \frac{1}{Pe_\lambda} \sim \frac{1}{Pe_\ell^{1/2}}, \tag{4.125}$$

$$\frac{\eta_\theta^C}{\ell} \sim \frac{1}{Pe_\eta^3} \sim \frac{1}{Pe_\ell^{3/4}}, \tag{4.126}$$

and

$$\frac{\eta_\theta^C}{\lambda_\theta} \sim \frac{1}{Pe_\eta} \sim \frac{1}{Pe_\lambda^{1/2}} \sim \frac{1}{Pe_\ell^{1/4}}, \tag{4.127}$$

or,

$$\boxed{Pe_\eta \sim Pe_\lambda^{1/2} \sim Pe_\ell^{1/4}}, \tag{4.128}$$

where $Pe = Pr Re$ is the Peclet number and, explicitly,

$$Pe_\eta = \frac{u\eta}{\alpha}, \quad Pe_\lambda = \frac{u\lambda}{\alpha}, \quad Pe_\ell = \frac{u\ell}{\alpha}.$$

Also, the ratio between η and η_θ^C gives

$$\frac{\eta}{\eta_\theta^C} \sim Pr^{3/4}, \tag{4.129}$$

and elimination of α^3/ϵ between Eqs. (4.122) and (4.124) yields

$$\left(\frac{\eta_\theta^C}{\lambda_\theta}\right)^2 \sim \frac{\lambda_\theta}{\ell}, \quad Pr \ll 1 \tag{4.130}$$

which may be used for another model for thermal intermittency.

Using the methodology proposed in Section 4.4, we obtained the Kolmogorov, Batchelor and Oboukhov-Corrsin scales. Also, we discovered a mesomicroscale for flows with $Pr \geq 1$. In the next section, following the same methodology, we obtain the scales of buoyancy-driven flows.

4.5.2 Buoyancy-Driven Flow

We directly proceed to the homogeneous flow (in which all averaged quantities except U_i and Θ are independent of position and, S_{ij} and $\partial\Theta/\partial x_i$ are constants) and the mean kinetic energy of turbulent fluctuations,

$$\mathcal{B} = \mathcal{P} + \epsilon, \tag{4.131}$$

where

$$\mathcal{B} = -g_i \overline{u_i \theta}/\Theta_0, \tag{4.132}$$

is the rate of buoyant energy production, g_i being vector acceleration of gravity and Θ_0 a characteristic temperature for isobaric ambient. The rate of thermal energy retains its earlier form,

$$\mathcal{P}_\theta = \epsilon_\theta. \tag{4.100}$$

On dimensional grounds, Eqs. (4.131) and (4.100) lead to

$$\mathcal{B} \sim \frac{u^3}{\ell} + \nu\frac{u^2}{\lambda^2} \tag{4.133}$$

and, using $u_\theta \sim u$ for buoyancy-driven flows,

$$u\frac{\theta^2}{\ell} \sim \alpha\frac{\theta^2}{\lambda_\theta^2} \, . \tag{4.134}$$

It seems reasonable to assume Eq. (4.20) to be a general property of flows and valid also under turbulent conditions. Accordingly, replace λ with λ_v and, furthermore, let $\lambda_v \sim \lambda_\theta$. Then, elimination of velocity between Eqs. (4.133) and (4.134) leads to a thermal Taylor scale,

$$\lambda_\theta \sim \ell^{1/3}\left(1 + \frac{1}{Pr}\right)^{1/6}\left(\frac{\nu\alpha^2}{B}\right)^{1/6} \, , \quad Pr \geq 1 \, , \tag{4.135}$$

or,

$$\lambda_\theta \sim \ell^{1/3}(1 + Pr)^{1/6}\left(\frac{\alpha^3}{B}\right)^{1/6} \, , \quad Pr \leq 1 \, , \tag{4.136}$$

where Eq. (4.135) explicitly includes the limit for $Pr \to \infty$ and is useful for fluids with $Pr \geq 1$, and Eq. (4.136) explicitly includes the limit for $Pr \to 0$ and is useful for fluids with $Pr \ll 1$.

For the isotropic limit of the homogeneous flow, letting

$$\left(\frac{\lambda_\theta}{\ell}\right) \to \eta_\theta \, , \tag{4.137}$$

Eqs. (4.135) and (4.136) are reduced to a thermal Kolmogorov scale introduced by Arpacı (1986, 1990a, 1995a,b),

$$\eta_\theta \sim \left(1 + \frac{1}{Pr}\right)^{1/4}\left(\frac{\nu\alpha^2}{B}\right)^{1/4} \, , \quad Pr \geq 1 \, , \tag{4.138}$$

or,

$$\eta_\theta \sim (1 + Pr)^{1/4}\left(\frac{\alpha^3}{B}\right)^{1/4} \, , \quad Pr \leq 1 \, . \tag{4.139}$$

Now, it is a simple matter to show from Eq. (4.138) that

$$\lim_{Pr\to\infty} \eta_\theta \to \left(\frac{\nu\alpha^2}{B}\right)^{1/4} \tag{4.140}$$

which, in view of

$$\lim_{Pr \to \infty} \mathcal{P} \to 0 \tag{4.141}$$

and

$$\mathcal{B} \to \epsilon , \tag{4.142}$$

leads to the Batchelor scale,

$$\lim_{Pr \to \infty} \eta_\theta \to \eta_\theta^B \sim \left(\frac{\nu \alpha^2}{\epsilon}\right)^{1/4} . \tag{4.143}$$

Also, from Eq. (4.139),

$$\lim_{Pr \to 0} \eta_\theta \to \left(\frac{\alpha^3}{\mathcal{B}}\right)^{1/4} \tag{4.144}$$

which, in view of

$$\lim_{Pr \to 0} \epsilon \to 0 , \tag{4.145}$$

$$\mathcal{B} \to \mathcal{P} , \tag{4.146}$$

and in a viscous layer order of magnitude thinner than η_θ,

$$\mathcal{P} \to \epsilon , \tag{4.147}$$

leads to the Oboukhov-Corrsin scale,

$$\lim_{Pr \to \infty} \eta_\theta \to \eta_\theta^C \sim \left(\frac{\alpha^3}{\epsilon}\right)^{1/4} . \tag{4.148}$$

Finally, for $Pr \sim 1$, because of the equipartition (on dimensional grounds) of buoyant production into inertial production and viscous dissipation, Eq. (4.131) becomes

$$\mathcal{B} \sim 2\epsilon , \tag{4.149}$$

and, both Eq. (4.138) and (4.139) lead to the Kolmogorov scale,

$$\lim_{Pr \to 1} \eta_\theta \to \eta \sim \left(\frac{\nu^3}{\epsilon}\right)^{1/4} . \tag{4.150}$$

The relation between the thermal scales and the integral scale may now be obtained by eliminating the factor $(1 + Pr^{-1})(\nu\alpha^2/\mathcal{B})$ between Eqs. (4.135) and (4.138) or Eqs. (4.136) and (4.139). Thus,

$$\left(\frac{\eta_\theta}{\lambda_\theta}\right)^2 = \frac{\lambda_\theta}{\ell} \tag{4.151}$$

which is identical to the relation obtained for the scales of forced flow for $Pr \ll 1$, and leads to a model for *thermal intermittency*.

Note that in the foregoing Kolmogorov scales given by Eqs. (4.138)-(4.139), $\mathcal{B}$ depends on velocity, and these scales cannot be in ultimate forms of the Kolmogorov scales for buoyancy driven flows. To eliminate velocity dependence, reconsider

$$\eta_\theta \sim \left(1 + \frac{1}{Pr}\right)^{1/4} \left(\frac{\nu\alpha^2}{\mathcal{B}}\right)^{1/4}, \tag{4.138}$$

and assume, on dimensional grounds,

$$\mathcal{B} \sim gu\theta/\Theta_0 . \tag{4.152}$$

Noting

$$\Theta_o^{-1} \sim \beta , \tag{4.153}$$

β being the coefficient of thermal expansion, rearrange Eq. (4.152) as

$$\mathcal{B} \sim gu\beta\theta , \tag{4.154}$$

or, with local velocity obtained from the limit of Eq. (4.134),

$$u \sim \alpha/\eta_\theta , \tag{4.155}$$

as

$$\mathcal{B} \sim g\alpha\beta\theta/\eta_\theta . \tag{4.156}$$

Insertion of Eq. (4.156) into Eq. (4.138) yields, after some rearrangement,

$$\eta_\theta \sim \left(1 + \frac{1}{Pr}\right)^{1/3} \left(\frac{\nu\alpha}{g\beta\theta}\right)^{1/3} . \tag{4.157}$$

Finally, assuming the rms of temperature fluctuations to be proportional to the imposed temperature difference (recall the argument following Eq. 4.76),

$$\theta \sim \Delta T \,, \tag{4.158}$$

$$\eta_\theta \sim \left(1 + \frac{1}{Pr}\right)^{1/3} \left(\frac{\nu\alpha}{g\beta\Delta T}\right)^{1/3} \,, \tag{4.159}$$

or, in terms of an integral scale ℓ,

$$\frac{\eta_\theta}{\ell} \sim \left(1 + \frac{1}{Pr}\right)^{1/3} Ra^{-1/3} \sim \Pi_N^{-1/3} \,, \tag{4.160}$$

where

$$Ra = \frac{g\beta\Delta T \ell^3}{\nu\alpha}$$

is the Rayleigh number.

In a similar manner, the thermal Taylor scale can be expressed in terms of Π_N. Recalling Eq. (4.151), now expressed as

$$\frac{\lambda_\theta}{\ell} \sim \left(\frac{\eta_\theta}{\ell}\right)^{2/3} \,, \tag{4.161}$$

we get from the combination of Eqs. (4.160) and (4.161),

$$\frac{\lambda_\theta}{\ell} \sim \Pi_N^{-2/9} \,. \tag{4.162}$$

Thus, we are able to express the Kolmogorov and Taylor thermal scales in terms of fundamental dimensionless number Π_N for buoyancy-driven flows developed in Chapter 2.

In some cases, buoyant flow is driven by internal energy generation rather than a temperature difference. Assume energy generated within volume $A\ell$, A being horizontal area and ℓ vertical height, be lost from A by conduction across a layer of thickness η_θ. From the balance between generated and lost energies,

$$u''' A\ell \sim kA\frac{\theta}{\eta_\theta} \,, \tag{4.163}$$

we have

$$\theta \sim \left(\frac{\eta_\theta \ell}{\alpha}\right) \Phi , \qquad (4.164)$$

where $\Phi = u'''/\rho c_p$. Now, buoyant production given by Eq. (4.156) becomes, in terms of Eq. (4.164),

$$\mathcal{B} \sim g\beta\ell\Phi , \qquad (4.165)$$

and the thermal Kolmogorov scale given by Eq. (4.138) becomes, in terms of Eq. (4.165),

$$\eta_\theta \sim \left(1 + \frac{1}{Pr}\right)^{1/4} \left(\frac{\nu\alpha^2}{g\beta\ell\Phi}\right)^{1/4} . \qquad (4.166)$$

Relative to vertical height ℓ,

$$\frac{\eta_\theta}{\ell} \sim \left(1 + \frac{1}{Pr}\right)^{1/4} Ra_I^{-1/4} , \qquad (4.167)$$

or,

$$\frac{\eta_\theta}{\ell} \sim \Pi_I^{-1/4} , \qquad (4.168)$$

where

$$\Pi_I \sim \frac{Ra_I}{1 + Pr^{-1}} = \frac{Ra_I Pr}{1 + Pr} \qquad (4.169)$$

and

$$Ra_I = \frac{g\beta\Phi\ell^5}{\nu\alpha^2} . \qquad (4.170)$$

Also, combining Eqs. (4.161) with (4.168), we get a thermal Taylor scale,

$$\frac{\lambda_\theta}{\ell} \sim \Pi_I^{-1/6} . \qquad (4.171)$$

Note that, we used subscript I to distinguish dimensionless numbers based on u''' from those based on ΔT.

The next chapter is devoted to heat transfer models for forced and buoyancy driven flows depending on the microscales developed in this chapter.

PROBLEMS

4-1) Show that the universal part of Eqs. (3.60) and (3.61),

$$\left(\frac{\nu^2}{b}\right)^{1/3},$$

is a Kolmogorov scale.

4-2) Show that the mean kinetic energy of turbulent fluctuations leads to Eq. (4.41).

4-3) Show that $\overline{\theta^2}$ of thermal fluctuations leads to Eq. (4.85).

4-4) What are Oboukhov-Corrsin's time and velocity scales? (Recall Prob. 1-1).

4-5) What are Batchelor's time and velocity scales? (Recall Prob. 1-4).

4-6) Determine the effect of Pr on the Oboukhov-Corrsin scale for $Pr < 1$.

4-7) Determine the microscales of rotating flows.

4-8) Determine the microscales of thermocapillary-driven flows.

CHAPTER 5

HEAT TRANSFER

In this chapter, we demonstrate microscale foundations of well-known heat transfer correlations. Also, we introduce models based on micro-scales, and correlate additional experimental data. The chapter begins with forced convection and terminates with natural convection.

5.1 FORCED CONVECTION

First, we recall from Chapter 3 the fact that any velocity gradient makes a layer of inviscid flow unstable (Kelvin-Helmholtz instability). Under the influence of viscosity, the layer becomes stable within the dissipation layer (of thickness η) over a boundary. Then, we extend the Tennekes model by assuming vortex tubes on boundaries to be stable and the ones above unstable (Fig. 5.1).

Now, we express the usual definitions of the coefficient of heat transfer and that of skin friction in terms of local dissipation scales, (u, η) for friction and (θ, η_θ) for heat transfer,

$$q_w \sim h\theta \sim k(\theta/\eta_\theta) \,, \tag{5.1}$$

$$\frac{1}{2}f \sim \frac{\tau_w}{\rho u^2} \sim \frac{\mu(u/\eta)}{\rho u^2} \sim \frac{\nu}{u\eta} \,, \tag{5.2}$$

rearrange Eqs. (5.1) and (5.2) as

$$h/k \sim \eta_\theta^{-1} \,, \tag{5.3}$$

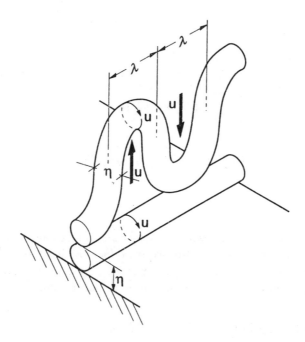

FIGURE 5.1: Stable and unstable vortex tubes next to a boundary.

$$u/\nu \sim \left(\frac{1}{2}f\right)^{-1} \eta^{-1} ,$$

(5.4)

establish their ratio,

$$\frac{h/k}{u/\nu} \sim \frac{1}{2}f\left(\frac{\eta}{\eta_\theta}\right) ,$$

(5.5)

and, nondimensionalizing the left-hand side of Eq. (5.5) by an integral scale, we obtain,

$$\left(\frac{Nu}{Re}\right)_\ell \sim \frac{1}{2}f\left(\frac{\eta}{\eta_\theta}\right) ,$$

(5.6)

where

$$Nu_\ell = h\ell/k \ \text{ and } \ Re_\ell = u\ell/\nu$$

are the Nusselt and Reynolds numbers. Next, we consider fluids with $Pr \geq 1$ and $Pr \ll 1$, respectively.

5.1.1 $Pr \geq 1$ (Gases, Liquids, Viscous Oils)

In the case of flow over a flat plate, the analogy between momentum and heat makes the ratio between Eqs. (5.3) and (5.4) given by Eqs. (5.5) and (5.6) an equality. Then, from Eq. (5.6), in terms of Eq. (4.116),

$$\left(\frac{Nu}{Re}\right)_\ell = \left(\frac{1}{2}f\right) Pr^{1/3} . \tag{5.7}$$

An alternative form of Eq. (5.7) expressed in terms of the Stanton number,

$$St = \frac{Nu}{RePr} , \tag{5.8}$$

is

$$StPr^{2/3} = \frac{1}{2}f , \quad Pr \geq 1 , \tag{5.9}$$

the well-known *Colburn correlation* relating heat transfer to skin friction. Foregoing development established this relation in terms of the microscales of turbulence rather than the usual approach which is the correlation of turbulent experimental data. Note that Eqs. (5.7) and (5.9) may also be obtained from (the similarity solution of the differential formulation or with approximate profiles satisfying the integral formulation of) laminar boundary layers. Consequently, they are *universal relations* valid for turbulent as well as laminar forced convection in fluids with $Pr \geq 1$. Only the explicitly stated friction coefficient distinguishes Eqs. (5.7) and (5.9) for turbulent flows from those for laminar flows.

The friction coefficient for turbulent flows is usually available from the correlation of experimental data. However, in tune with the philosophy of present monograph, a model based on the local dissipation scales is constructed here. Recall and rearrange Eq. (5.2) as

$$\frac{1}{2}f \sim Re_\eta^{-1} , \tag{5.10}$$

where

$$Re_\eta = u\eta/\nu . \tag{5.11}$$

Then, in view of Eq. (4.66),

$$\frac{1}{2}f \sim Re_\ell^{-1/4} . \tag{5.12}$$

Now, the combination of Eqs. (5.7) and (5.12) yields

$$Nu_\ell \sim Re_\ell^{3/4} Pr^{1/3} , \quad Pr \geq 1 . \tag{5.13}$$

The experimental data confirms the validity of this relation for turbulent heat transfer over a flat plate and in a pipe, provided Reynolds number remains moderate in the case of latter.

An interesting alternative development leading directly to Eq. (5.13) is worth mentioning here. In terms of the thermal intermittency given by Eq. (4.120), the Nusselt number yields

$$Nu_\ell \sim \frac{\ell}{\eta_\theta} \sim \left(\frac{\ell}{\lambda_\theta}\right)^{3/2} Pr^{-1/6} \tag{5.14}$$

or, after some rearrangement,

$$Nu_\ell \sim \left(\frac{\ell}{\lambda}\right)^{3/2} \left(\frac{\lambda}{\lambda_\theta}\right)^{3/2} Pr^{-1/6} \tag{5.15}$$

which, in view of Eqs. (4.63) and (4.108), leads to Eq. (5.13).

A second model takes into account a core effect above the viscous sublayer. Now, Eq. (4.74) is replaced by

$$U \sim u + u_c , \tag{5.16}$$

subscript c denoting the core effect. For the sublayer, noting $u \sim \nu/\eta$, let

$$\tau \sim \mu \frac{u}{\eta} \sim \rho u^2 \tag{5.17}$$

which states a same order of magnitude contribution from viscous and Reynolds stresses. For the core, neglecting the viscous effect, let

$$\tau_c \sim \rho u_c^2 . \tag{5.18}$$

Then, assuming

$$\tau \sim \tau_c , \tag{5.19}$$

and using Eqs. (5.17) and (5.18), rearrange Eq. (5.16) as

$$\tau \sim \frac{\mu \dfrac{U}{\eta}}{\left(1 + \dfrac{\nu}{\eta u_c}\right)} . \tag{5.20}$$

Next, consider Eq. (4.77) for u_c, note

$$\frac{u_c \ell}{\nu} \sim \left(\frac{\ell}{\lambda}\right)^2 , \qquad (5.21)$$

and rearrange $\nu/\eta u_c$ of Eq. (5.20) as

$$\frac{\nu}{\eta u_c} \sim \frac{\nu}{u_c \ell}\left(\frac{\ell}{\eta}\right) \sim \left(\frac{\lambda}{\ell}\right)^2 \frac{\ell}{\eta} . \qquad (5.22)$$

Thus,

$$\tau \sim \frac{\mu \dfrac{U}{\eta}}{1 + \left(\dfrac{\lambda}{\ell}\right)^2 \dfrac{\ell}{\eta}} , \qquad (5.23)$$

or, in terms of Eqs. (4.63) and (4.64),

$$\tau \sim \frac{\mu \dfrac{U}{\eta}}{1 + \dfrac{1}{Re_\ell^{1/4}}} . \qquad (5.24)$$

Next, we repeat the steps between Eqs. (5.16) and (5.24) for thermal energy. First, replace Eq. (4.158) by

$$\Delta T \sim \theta + \theta_c , \qquad (5.25)$$

subscript c denoting the core effect. For the *thermal sublayer*, noting $u_\theta \sim \alpha/\eta_\theta$, let

$$q \sim k\frac{\theta}{\eta_\theta} \sim \rho c_p u_\theta \theta \qquad (5.26)$$

which states a same order of magnitude for conduction and Reynolds flux. For the *thermal core*,

$$q_c \sim \rho c_p (u_\theta)_c \theta_c . \qquad (5.27)$$

Then, assuming

$$q \sim q_c \qquad (5.28)$$

and using Eqs. (5.26) and (5.27), rearrange Eq. (5.25) as

$$q \sim \frac{k\dfrac{\Delta T}{\eta_\theta}}{1 + \dfrac{\alpha}{(u_\theta)_c \eta_\theta}} \; . \qquad (5.29)$$

For the denominator, consider

$$\frac{(u_\theta)_c \eta_\theta}{\alpha} = \frac{u_c \ell}{\nu} \frac{\eta}{\ell} \frac{\eta_\theta}{\eta} \frac{(u_\theta)_c}{u_c} \frac{\nu}{\alpha}$$

or, in terms of Eqs. (5.21), (4.63), (4.64), (4.116), (4.104) for core, and (4.108),

$$\frac{(u_\theta)_c \eta_\theta}{\alpha} \sim Re_\ell^{1/4} Pr^{1/3} \; . \qquad (5.30)$$

Thus, Eq. (5.29) becomes, in terms of Eq. (5.30),

$$q \sim \frac{k\dfrac{\Delta T}{\eta_\theta}}{1 + \dfrac{1}{Re_\ell^{1/4} Pr^{1/3}}} \; . \qquad (5.31)$$

Now, back to definitions of the coefficient of heat transfer and skin friction in terms of Eqs. (5.24) and (5.31), and replace Eqs. (5.3) and (5.4) with

$$\frac{h}{k} \sim \frac{\eta_\theta^{-1}}{1 + \dfrac{1}{Re_\ell^{1/4} Pr^{1/3}}} \qquad (5.32)$$

and

$$\frac{U}{\nu} \sim \frac{\eta^{-1}}{1 + \dfrac{1}{Re_\ell^{1/4}}} \; . \qquad (5.33)$$

Then, from the ratio between Eqs. (5.32) and (5.33), noting $(h/k)/(U/\nu) = (Nu/Re)_\ell$, recalling the analogy between momentum and heat, we have

$$\left(\frac{Nu}{Re}\right)_\ell = \left(\frac{1}{2}f\right)\left(\frac{\eta}{\eta_\theta}\right) \Phi(Re_\ell, Pr) \; , \quad Pr \geq 1 \; , \qquad (5.34)$$

or, in terms of Eqs. (4.116) and (5.8),

$$StPr^{2/3} = \left(\frac{1}{2}f\right) \Phi(Re_\ell, Pr), \quad Pr \geq 1, \tag{5.35}$$

where

$$\Phi = \frac{1 + \dfrac{C}{Re_\ell^{1/4}}}{1 + \dfrac{C}{Re_\ell^{1/4}Pr^{1/3}}} \tag{5.36}$$

is the core effect, C being a constant depending on the structure. Note that, for $Re_\ell \to \infty$, the core effect diminishes, as expected.

With an approximate value of $C \approx 5$, obtained somewhat loosely by fitting Eq. (5.35) for $Re_\ell = 3 \times 10^4$ and $Pr = 100$ to the empirical correlation of Ribaud (1941),

$$\frac{St}{\frac{1}{2}f} = \frac{1}{1 + 0.75(Pr^{2/3} - 1)}.$$

In Fig. 5.2, we compare the present model with some earlier models based on eddy diffusivity and the law of the wall. The figure includes also a three-layer model of Karman.

Next, we proceed to a model for liquid metals.

5.1.2 $Pr \ll 1$ (Liquid Metals)

In this case, Eq. (5.6) yields, in terms of Eqs. (5.12) and (4.129),

$$Nu_\ell \sim Pe_\ell^{3/4}, \quad Pr \to 0. \tag{5.37}$$

An alternative is the use of Nusselt number,

$$Nu \sim \frac{\ell}{\eta_\theta^C}, \tag{5.38}$$

which, in view of Eq. (4.126), directly leads to Eq. (5.37). A two layer model including Prandtl effect for liquid metals is left to the reader (see, Prob. 5-1)

Because of wide scatter in experimental literature, there are a number of correlations on liquid metal heat transfer. Equation (5.37) is

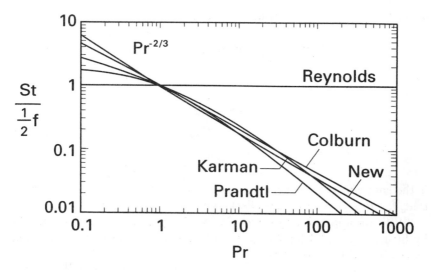

FIGURE 5.2: Comparison between various models for heat transfer correlation.

consistent with some of these correlations. However, new data is needed for further comparison. This data should distinguish the surface wetting and nonwetting liquid metals, should explicitly include the effect of Prandtl number, and should be correlated with an improved model depending on Pr.

5.2   NATURAL CONVECTION

Here, we consider two cases, the vertical plate and the classical Rayleigh Benard problem.

5.2.1   Vertical Plate

A mathematical model by Churchill and Chu (1975) correlates the extensive laminar and turbulent data with one correlation,

$$Nu^{1/2} = 0.825 + \frac{0.387 Ra^{1/6}}{[1 + (0.492/Pr)^{9/16}]^{8/27}} . \tag{5.39}$$

Since each regime is fundamentally different than the other, Eq. (5.39) may be considered as an approximate representation of both regimes rather than being accurate for each one.

A model based on microscales, obtained from the inverse of Eq. (4.160), is

$$Nu \sim \frac{\ell}{\eta_\theta} \sim \Pi_N^{1/3} , \tag{5.40}$$

, or, explicitly,

$$Nu = C_1 \Pi_N^{1/3} , \quad \Pi_N = \frac{Ra}{1 + C_0 Pr^{-1}} . \tag{5.41}$$

Using a least-squares fitting, Arpacı and Kao (1997) obtain $C_0 = 0.164$ and $C_1 = 0.150$ for the turbulent regime of Eq. (5.39) or, explicitly, for $10^9 \leq \Pi_N \leq 10^{12}$,

$$Nu = 0.15\Pi_N^{1/3} , \quad \Pi_N = \frac{Ra}{1 + 0.164 Pr^{-1}} . \tag{5.42}$$

Between the two constants, C_0 is extremely sensitive to data. The numerical value of C_1 is expected to be quite accurate but that of C_0 needs further examination.

Next, we consider the classical Rayleigh-Benard problem of buoyancy-driven flow between tow horizontal plates.

5.2.2   Rayleigh-Benard Problem

Assume the temperature difference between the plates is large and fully developed turbulent conditions prevail. This is an ideal problem for testing models based on microscales because of the wealth of experimental, analytical and computational literature.

In a manner similar to the two-layer turbulence model of Prandtl and Taylor for forced convection, let the buoyancy driven turbulent flow be described by a sublayer next to each plate and a core between these layers. Assume each sublayer be characterized by local dissipation scales and the core by volume-averaged dissipation scales.

The mean heat flux in the sublayer, in terms of local dissipation and $u \sim \alpha/\eta_\theta$ or $\rho c_p u \sim k/\eta_\theta$, is

$$q \sim k\frac{\theta}{\eta_\theta} \sim \rho c_p u\theta \tag{5.43}$$

which shows a same order of magnitude conduction and convection. The heat flux in the core, after neglecting the small effect of conduction, is

$$q_c \sim \rho c_p u_c \theta_c , \tag{5.44}$$

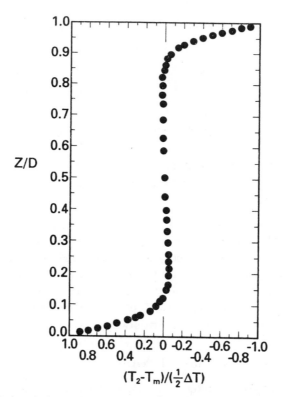

FIGURE 5.3: A typical mean temperature profile in a water layer with a gradient reversal. $Ra = 3.97 \times 10^6$, $D = 3.8$ cm, $\Delta T = 0.441$ °C. Chu and Goldstein (1973).

where the subscript c indicates core. At the interface between sublayer and core

$$q \sim q_c . \qquad (5.45)$$

In view of the temperature reversal in the core (Fig. 5.3), demonstrated experimentally by Thomas and Townsend (1957), Gill (1967) and Chu and Goldstein (1973), and numerically by Herring (1963) and Elder (1969), let

$$\theta \sim \Delta T + \theta_c , \qquad (5.46)$$

where ΔT is the imposed temperature difference across the plates. In-serting θ of Eq. (5.43) and θ_c of Eq. (5.44) into Eq. (5.46), noting

Eq. (5.45),

$$q(1 - \alpha/u_c\eta_\theta) \sim k\Delta T/\eta_\theta \tag{5.47}$$

which, in terms of the Nusselt number,

$$Nu = \frac{q}{k(\Delta T/\ell)} , \tag{5.48}$$

leads to

$$Nu \sim \frac{\ell/\eta_\theta}{1 - (\ell/\eta_\theta)(u_c\ell/\alpha)^{-1}} , \tag{5.49}$$

where the numerator ℓ/η_θ shows the contribution of sublayer and $(\ell/\eta_\theta)(u_c\ell/\alpha)^{-1}$ that of core to heat transfer. Also, for Eq. (5.49) in length scales alone, consider the velocity obtained from Eq. (4.103) by replacing u_θ with u,

$$u_c \sim \alpha\frac{\ell}{\lambda_\theta^2} , \tag{5.50}$$

or, its equivalent,

$$\frac{u_c\ell}{\alpha} \sim \left(\frac{\ell}{\lambda_\theta}\right)^2 . \tag{5.51}$$

With this relation, Eq. (5.49) becomes

$$Nu \sim \frac{\ell/\eta_\theta}{1 - (\ell/\eta_\theta)(\ell/\lambda_\theta)^{-2}} \tag{5.52}$$

and, in view of Eqs. (4.160) and (4.162), yields

$$Nu \sim \frac{\Pi_N^{1/3}}{1 - \Pi_N^{-1/9}} \tag{5.53}$$

which may now be converted to an equality by three constants C_0, C_1, and C_2,

$$Nu = \frac{C_1\Pi_N^{1/3}}{1 - C_2\Pi_N^{-1/9}} , \tag{5.54}$$

where

$$\Pi_N = \frac{Ra}{1 + C_0 Pr^{-1}} = \left(\frac{Pr}{C_0 + Pr}\right) Ra \ . \tag{5.55}$$

Equation (5.54) provides a heat transfer correlation for natural convection between two parallel plates. This correlation applies for fluids of any Prandtl number and Rayleigh number under turbulent conditions. Although the numerical values of the constants need to be determined from experimental data, they are expected to be universal.

Arpacı and Dec (1987) computed these values by least-squares fitting of Eq. (5.54) to experimental data from literature. The available experimental data and the selection of the best liquid metal, air and water data for computation of the constants is discussed in the following literature review. Silveston (1958) covers a range of high Prandtl number liquids, Globe and Dropkin (1959) cover liquid metals, water and viscous oils. However, these data except that of liquid metals ($Pr \approx 0.02$) have been superseded by later work. Since the value of C_0 requires low Prandtl data, the liquid metal data of Globe and Dropkin is used. Gases ($Pr \approx 0.7$) are covered by Threlfall (1975), Fitzjarrald (1976), and Goldstein and Chu (1969). All these authors show plotted data rather than tabulated values. The data of Goldstein and Chu has the least scatter and some tabulated values, so it is the air data used. Water ($Pr \approx 6$) is covered by Chu and Goldstein (1973), Garon and Goldstein (1973), and Goldstein and Tokuda (1980). The data of Garon and Goldstein is tabulated and covers the widest range of Rayleigh numbers, so it is the water data used. Viscous oil data ($Pr > 6$) is omitted from computations for two reasons. There is a lack of high quality recent data, and the constants are not sensitive to Prandtl effects for $Pr \gg 1$. The only recent data for viscous oils is that of Somerscales and Gazda (1969), whose data are not very extensive and differ considerably from that of Goldstein and others.

Before computing the constants, the following modifications were made to the selected data. The data of Globe and Dropkin for $Ra < 10^6$ were omitted since they are likely not to be fully turbulent. Also, the data of Goldstein and Chu which has an aspect ration (width/height) of one are omitted since they cannot be considered free of edge effects. From a least-squares fit of the selected data, the numerical constants are found by Arpacı and Dec (1987) to be: $C_0 = 0.04140$, $C_1 = 0.04707$ and $C_2 = 1.734$. These values are accurate for the data used; however,

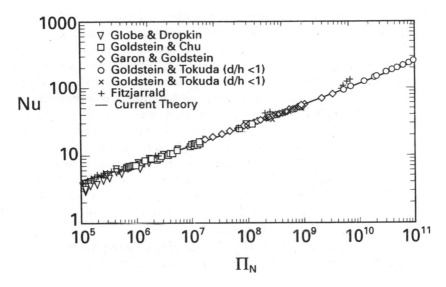

FIGURE 5.4: A comparison of proposed model with experimental literature for $0.02 \leq Pr \leq 6$ and $10^6 \leq Ra \leq 10^{11}$.

improved future data may lead to refinement. This is especially true for C_0, since its value relies heavily on low Prandtl number (liquid metal) data, which is quite old and does not extend beyond $Ra = 3.4 \times 10^7$.

Figure 5.4 shows the correlation of the experimental data with the proposed model. In addition to the data used to compute constants, the data of Fitzjarrald, and Goldstein and Tokuda as well as the omitted points of Globe and Dropkin, and Goldstein and Chu are plotted. The omitted low Ra (and low Π_N) data of Globe and Dropkin and the low aspect ratio data of Goldstein and Tokuda (high Π_N) deviate from the curve. It can be seen that, when converted into terms of Π_N, data from fluids of all Prandtl numbers collapses on to a single curve. It should be noted that the plot of the proposed model in Fig. 5.4 is not a straight line due to core effect. This fact has been noticed by several previous workers who interpreted it in terms of a series of discrete transitions. A review and discussion of these proposed transitions may be found in Chu and Goldstein (1973). It is important to note that the discussion is limited to proposed transitions in the fully turbulent range, $Ra > 10^6$. They should not be confused with the transitions for $Ra < 10^6$, which Krishnamurti (1970a,b, 1973) has shown, through flow visualization techniques, to be due to changing flow patterns. Krishnamurti's experiments are impressively complemented by analytical and computational studies of Busse and co-workers, summarized in Busse (1981). Four transitions in fully turbulent regime ($Ra > 10^6$) are explicitly reported by Garon and Goldstein (1973), although they note that "their validity is open to question." Their plot for these transitions is reproduced in Fig. 5.5. The model developed in this paper offers an alternative interpretation of these data. In the model, curvature resulting from the core effect produces a smooth curve through these apparent transitions. This may be seen in Fig. 5.6, which presents the proposed model and the data and transitions of Garon and Goldstein, plotted in the same linear fashion as Fig. 5.5. The linear plotting dramatizes the core effect noted in Fig. 5.6. The continuous curve produced by the present model may be seen to fit the data very well.

An earlier model by Long (1976) also refutes these proposed transitions in the fully turbulent regime. The Long model leads to

$$Nu = \frac{Ra^{1/3}Pr^{2/3}}{\left[A_0\delta + 2C\Delta - \gamma(NuRa)^{-s/4}\right]^{4/3}} \, , \tag{5.58}$$

which explicitly includes the core effect. Here A_0, δ, C, δ and γ are functions of Pr and s is a constant exponent. Equation (5.58) ultimately reduces to

$$Nu = \frac{C_1 Ra^{1/3}}{\left[1 - C_2(NuRa)^{-s/4}\right]^{4/3}} \, , \tag{5.59}$$

where the constants are functions of Prandtl number, so different numerical values must be calculated for each fluid. In contrast, the present model is valid for any Prandtl number, as well as for any Rayleigh number in the turbulent range. It is general in that the parameters involved with the theory are numerical constants independent of the Prandtl number.

For buoyant flow driven by energy generation rather that a temperature difference, inserting Eqs. (4.168) and (4.171) into Eq. (5.52), we obtain

$$Nu \sim \frac{\Pi_I^{1/4}}{1 - \Pi_I^{-1/12}} \, . \tag{5.60}$$

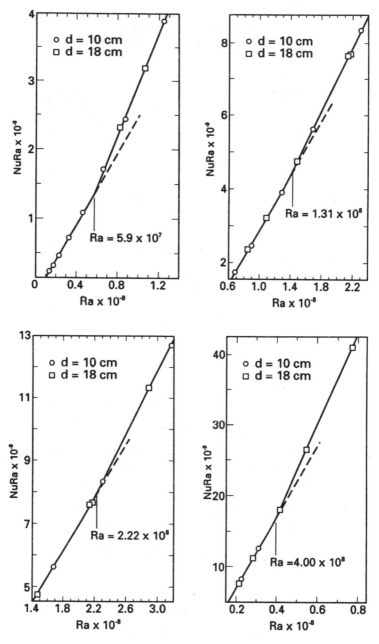

FIGURE 5.5: "Four transitions" (Garon and Goldstein, 1973).

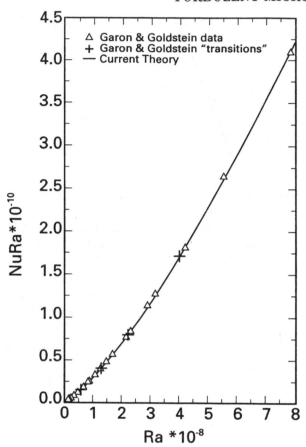

FIGURE 5.6: A comparison of proposed model with experimental data about four "transitions" plotted in Garon and Goldstein (1973).

The two limits of this result,

$$\lim_{Pr\to 0} Nu \sim \frac{(Pr\,Ra_I)^{1/4}}{1 - (Pr\,Ra_I)^{-1/12}} \tag{5.56}$$

and

$$\lim_{Pr\to\infty} Nu \sim \frac{Ra_I^{1/4}}{1 - Ra_I^{-1/12}}\,, \tag{5.57}$$

are identical to the models already proposed by Cheung (1980). Thus, the present study generalizes, via microscales introduced for buoyancy driven flows, two Cheung correlations into Eq. (5.60) which is valid for fluids of any Prandtl number. Now, Eq. (5.60) may be written as an equality in terms of three constants

$$Nu = \frac{C_1 \Pi_I^{1/4}}{1 - C_2 \Pi_I^{-1/12}}, \quad \Pi_I = \left(\frac{Pr}{C_0 + Pr}\right) Ra_I. \quad (5.61)$$

Equation (5.61) provides a heat transfer correlation for turbulent natural convection driven by energy generation between two parallel plates.

The experimental literature on the buoyant turbulent flow driven by volumetric internal energy generation is confined to the studies of Tritton and Zarraga (1967), Fiedler and Wille (1971), Kulacki and Nagle (1975), and Kulacki and Emara (1977). These studies employ electrolytically heated water for which Pr remains within the narrow range of 6-7. If one assumes $C_0 \ll 1$ indicating to a small inertial effect (see Arpacı, 1990a), the numerical value of

$$\left(\frac{Pr}{C_0 + Pr}\right)^{1/4}$$

can be very closely approximated by unity. Then

$$\Pi_I \to Ra_I, \quad Pr > 1$$

and Nu given by Eq. (5.61) is reduced to

$$Nu = \frac{C_1 Ra_I^{1/4}}{1 - C_2 Ra_I^{-1/12}}. \quad (5.62)$$

Cheung employs the data of Kulacki and Emara and proposes

$$Nu = \frac{0.206 Ra_I^{1/4}}{1 - 0.847 Ra_I^{-1/12}}. \quad (5.63)$$

Figure 5.7 taken from Cheung (1980) shows the correlation of the experimental data by Eq. (5.63). A correlation for any Prandtl number incorporating the numerical values of C_0, C_1 and C_2 into Eq. (5.61) needs data for another Prandtl range (preferably for liquid metals) which is not presently available.

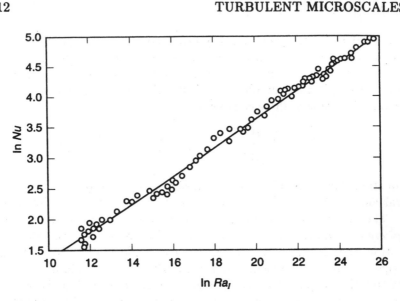

FIGURE 5.7: ln Nu versus ln Ra_I. —, Cheung model given by Eq. (5.63); o , Data of
Kulacki and Emara (1977).

PROBLEMS

5-1) Determine the effect of Pr on Nu for $Pr \ll 1$.

5-2) Determine the microscales of buoyancy-driven flows for a specified q_w.

5-3) Develop a heat transfer model for rotating turbulent flows.

5-4) Develop a heat transfer model for thermocapillary-driven turbulent flows.

CHAPTER 6

MASS TRANSFER

In this chapter we use reacting flows as an illustration for mass transfer and specifically consider diffusion flames. Our objective is to introduce microscales for turbulent flames, to construct models for fuel consumption, and to correlate experimental data. We begin with forced flames and, subsequently proceed to buoyancy-driven flames.

6.1 FORCED FLAMES

First, consider a dimensional interpretation of laminar flames which proves convenient in Section 6.1.2 on turbulent flames.

6.1.1 Laminar Flame

Reconsider the first part of Spalding (1954) which deals with diffusion flame over a flat plate (Fig 6.1). The balance of momentum integrated over momentum boundary layer thickness δ is

$$\rho U^2 \frac{d}{dx} \int_0^\delta \left(1 - \frac{u}{U}\right)\left(\frac{u}{U}\right) dy - \rho U v_w = \mu \left(\frac{\partial u}{\partial y}\right)_w, \qquad (6.1)$$

ρ being the density, μ the dynamic viscosity, u the longitudinal velocity, U its free stream value, and v_w the velocity normal to the fuel surface. Let $Le = \alpha/D = 1$, Le being the Lewis number, α and D the thermal and mass diffusivities, respectively. Then, the balance of

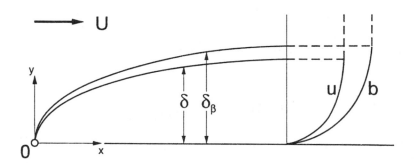

FIGURE 6.1: Forced diffusion flame over a flat plate.

generalized diffusion property b, the Schvab-Zeldovich (heat+oxidizer) property integrated over its boundary layer thickness δ_β, gives

$$\rho U B \frac{d}{dx} \int_0^{\delta_\beta} \frac{u}{U} \left(\frac{b_\infty - b}{B} \right) dy - \rho B v_w = \rho D \left(\frac{\partial b}{\partial y} \right)_w , \qquad (6.2)$$

where b and the transfer number B (Busemann, 1993) are defined as,

$$b = (Y_O Q / \nu_O M_O + h) / h_{fg} , \qquad (6.3)$$

and

$$B = b_\infty - b_w , \qquad (6.4)$$

or, in terms of Eq. (6.3), explicitly,

$$B = (Y_{O\infty} Q / \nu_O M_O - h_w) / h_{fg} . \qquad (6.5)$$

Here, Y_O is the mass fraction of the oxidizer; $Y_{O\infty}$ its ambient value; Q is the heat released according to a single global chemical reaction

$$\nu_F \text{ (Fuel)} + \nu_O \text{ (Oxidant)} \rightarrow \text{Products} + Q \text{ (heat)} , \qquad (6.6)$$

where

$$\frac{Q}{\nu_O M_O} = \left(\frac{Q}{\nu_F M_F} \right) \left(\frac{\nu_F M_F}{\nu_O M_O} \right) \qquad (6.7)$$

is the heat release per unit mass of oxidizer, $Q/\nu_F M_F$ the lower heating value (heat release per kg of fuel), $\nu_F M_F/\nu_O M_O$ the stoichiometric fuel to oxidant ratio (kg fuel/kg oxidant), ν_F, ν_O, and M_F, M_O being the fuel and oxidant stoichiometric coefficients and molecular weights, respectively, h the specific enthalpy of gas relative to ambient temperature; $h_w = c_p(T_w - T_\infty)$, c_p the specific heat of gas, T_w and T_∞ fuel surface and ambient temperatures, respectively, and h_{fg} the heat of evaporation. The local mass transfer (burning rate per unit area) is

$$M'' = \rho v_w = \rho D \left(\frac{\partial b}{\partial y}\right)_w , \qquad (6.8)$$

and the mass transfer averaged over a length ℓ is

$$M' = \int_0^\ell M'' \, dx . \qquad (6.9)$$

Now, consider a dimensional interpretation of the foregoing formulation.

Following Arpacı and Li (1995), the balance of momentum given by Eq. (6.1) becomes,

$$U\frac{U}{\ell} \sim v_w\frac{U}{\delta} + \nu\frac{U}{\delta^2} , \qquad (6.10)$$

and the balance of b-property given by Eq. (6.2) becomes,

$$U_\beta\frac{B}{\ell} \sim v_w\frac{B}{\delta_\beta} + D\frac{B}{\delta_\beta^2} . \qquad (6.11)$$

In terms of the mass balance on the fuel surface,

$$\rho v_w \sim \rho D\frac{B}{\delta_\beta} . \qquad (6.12)$$

Equations (6.10) and (6.11) may be rearranged as

$$U\frac{U}{\ell} \sim \nu \left(1 + \frac{\delta}{\delta_\beta}\frac{B}{\sigma}\right)\frac{U}{\delta^2} \qquad (6.13)$$

and

$$U_\beta\frac{B}{\ell} \sim D(1 + B)\frac{B}{\delta_\beta^2} . \qquad (6.14)$$

For notational convenience, let

$$\nu_\beta = \nu(1 + B') , \tag{6.15}$$

$$D_\beta = D(1 + B) , \tag{6.16}$$

where

$$B' = \chi\frac{B}{\sigma} , \quad \chi = \frac{\delta}{\delta_\beta} , \tag{6.17}$$

$\sigma = \nu/D$ being the Schmidt number. Then, Eqs. (6.13) and (6.14) are reduced to

$$U\frac{U}{\ell} \sim \nu_\beta\frac{U}{\delta^2} \tag{6.18}$$

and

$$U_\beta\frac{B}{\ell} \sim D_\beta\frac{B}{\delta_\beta^2} . \tag{6.19}$$

In view of the similarity between Eqs. (6.18) and (6.19),

$$U_\beta = U\frac{\delta_\beta}{\delta} , \tag{6.20}$$

and

$$\chi = \sigma_\beta^{1/3} , \tag{6.21}$$

σ_β being the flame Schmidt number,

$$\sigma_\beta = \frac{\nu_\beta}{D_\beta} = \left(\frac{\nu_\beta}{\nu}\right)\left(\frac{D}{D_\beta}\right)\left(\frac{\nu}{D}\right) = \left(\frac{1 + B'}{1 + B}\right)\sigma . \tag{6.22}$$

Equations (6.20) and (6.21) are strictly valid when $\chi \geq 1$ which is not the case here. However, χ remains near unity and the error introduced by this assumption remains small.

From Eq. (6.18)

$$\frac{\ell}{\delta} \sim Re_\beta^{1/2} , \tag{6.23}$$

where

$$Re_\beta = \frac{U\ell}{\nu_\beta} = \left(\frac{\nu}{\nu_\beta}\right)\frac{U\ell}{\nu} = \frac{Re}{(1+B')} , \qquad (6.24)$$

$Re = U\ell/\nu$ being the Reynolds number. Now, the burning rate over a length ℓ,

$$\frac{M'}{\rho D} \sim B\frac{\ell}{\delta_\beta} = \chi B\left(\frac{\ell}{\delta}\right) , \qquad (6.25)$$

becomes, in terms of Eqs. (6.21) and (6.23),

$$\frac{M'}{\rho D} \sim BRe_\beta^{1/2}\sigma_\beta^{1/3} , \qquad (6.26)$$

or, explicitly,

$$\frac{M'}{\rho D} \sim \frac{BRe^{1/2}}{(1+B')^{1/2}}\left(\frac{1+B'}{1+B}\right)^{1/3}\sigma^{1/3} , \qquad (6.27)$$

or,

$$\frac{M'}{\rho D Re^{1/2}\sigma^{1/3}} \sim \frac{B}{(1+B')^{1/6}(1+B)^{1/3}} . \qquad (6.28)$$

For $\chi \sim 1$ and $\sigma \sim 1$, $B \sim B'$ and Eq. (6.28) is reduced, with a parameter C, to

$$\frac{M'}{\mu Re^{1/2}} = \frac{CB}{(1+B)^{1/2}} \qquad (6.29)$$

which is identical, in the sense of explicit B-dependence, to

$$\frac{M'}{\mu Re^{1/2}} = B\left[\frac{a_1(\delta_m/\delta)}{2(1+B)}\right]^{1/2} , \qquad (6.30)$$

the result obtained by Spalding (1954), a_1 being the coefficient of the linear term in the assumed cubic velocity profile, δ_m the momentum thickness and δ the boundary layer thickness. From Eqs. (6.29) and (6.30),

$$C = \left[\frac{a_1}{2}\left(\frac{\delta_m}{\delta}\right)\right]^{1/2} . \qquad (6.31)$$

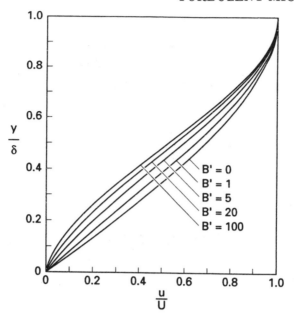

FIGURE 6.2: B' effect on velocity.

An inspection of Spalding reveals that δ_m/δ remains almost constant but a_1 decays monotonically and is mainly responsible for the B-dependence of C. Also, the velocity assumes a curvature reversal (inflection) depending on B (Fig. 6.2). The foregoing dimensional considerations need to be complemented by including this dependence into Eq. (6.29). Spalding includes the boundary condition,

$$\rho\nu\frac{db}{dy} = \rho D\frac{d^2b}{dy^2} \,, \quad y = 0 \,, \tag{6.32}$$

whose effect is missing from these considerations. Now, assume Eq. (6.32) to apply for $0 \leq y \leq \delta$, and recover from twice integration the burning rate obtained from the stagnant film theory,

$$\frac{M'}{\rho D} = \chi\ln(1 + B)\frac{\ell}{\delta} \,. \tag{6.33}$$

This suggests a somewhat logarithmic rather than linear behavior for the numerator of Eq. (6.29). It is easy to show, by an expansion and inversion, that

$$\ln(1 + B) = \frac{B}{1 + \dfrac{B}{2} - \dfrac{B^2}{12} + \cdots} \tag{6.34}$$

which can be simulated by

$$\frac{B}{(1 + C_0 B)^n} . \tag{6.35}$$

For $n < 1$, the denominator of Eq. (6.35) is an alternating series like the one in the denominator of Eq. (6.34). In terms of

$$C = \frac{C_1}{(1 + C_0 B)^n} , \tag{6.36}$$

the burning rate becomes now

$$\frac{M'}{\mu Re^{1/2}} = \frac{C_1 B}{(1 + C_0 B)^n (1 + B)^{1/2}} . \tag{6.37}$$

For $B = 0$, the heat transfer literature based on cubic profiles gives

$$C_1 = 0.646 .$$

For $B \to \infty$, Spalding approaches $B^{1/4}$ which requires

$$n = \frac{1}{4} .$$

Thus,

$$\frac{M'}{\mu Re^{1/2}} = \frac{0.646 B}{(1 + C_0 B)^{1/4} (1 + B)^{1/2}} . \tag{6.38}$$

An inspection of the literature (Kanury, 1977) reveals an approximate upper bound of $B \sim 10$ for liquid fuels. A least squares fit of Eq. (6.38) to Spalding for $B \leq 20$ yields

$$C_0 = 0.525 ,$$

and

$$\frac{M'}{\mu Re^{1/2}} = \frac{0.646 B}{(1 + 0.525 B)^{1/4} (1 + B)^{1/2}} . \tag{6.39}$$

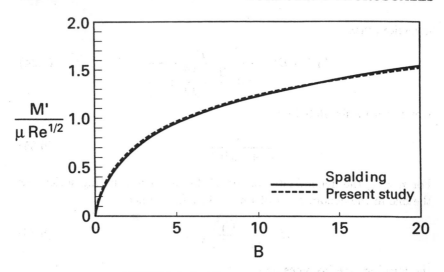

FIGURE 6.3: Spalding versus present study.

Fig. 6.3 shows the comparison of Eq. (6.39) with Spalding against $B \leq$ 20.

For reasons to be clear in the next section, consider also the exact solution obtained by Emmons (1956). For $B = 0$, the literature based on the similarity solution gives

$$C_1 = 0.664 .$$

For $B \to \infty$, Emmons approaches a constant which requires

$$n = \frac{1}{2} .$$

Thus

$$\frac{M'}{\mu Re^{1/2}} = \frac{0.664B}{(1 + C_0 B)^{1/2}(1 + B)^{1/2}} . \qquad (6.40)$$

A least squares fit of Eq. (6.40) to Emmons for $B \leq 20$ yields

$$C_0 = 0.354$$

and

$$\frac{M'}{\mu Re^{1/2}} = \frac{0.664B}{(1 + 0.354B)^{1/2}(1 + B)^{1/2}} . \qquad (6.41)$$

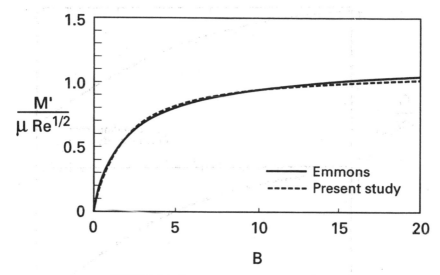

FIGURE 6.4: Emmons versus present study.

Fig. 6.4 shows the comparison of Eq. (6.41) with Emmons against $B \leq$ 20.

Although the main interest lies in the fuel consumption, for convenience of the next section on turbulent diffusion flames, consider also the heat transfer and skin friction. As is well-known, for $B \sim B'$ and $\chi \sim 1$, the similarity leads to

$$\lim_{B \to 0} \left(\frac{M'}{\mu B} \right) = Nu = C_f Re \,, \tag{6.42}$$

or

$$\frac{Nu}{Nu_{B=0}} = \frac{C_f}{(C_f)_{B=0}} = \frac{1}{(1 + C_0 B)^n (1 + B)^{1/2}} \,, \tag{6.43}$$

N being the Nusselt number and C_f the drag coefficient, subscript $B = 0$ corresponding to the case without boundary mass transfer. The values for n and C_0 are those assumed for the fuel consumption. Fig. 6.5 shows $Nu/Nu_{B=0} = C_f/(C_f)_{B=0}$ versus $B \leq 20$.

The foregoing dimensional arguments are now extended to turbulent flames in terms of the general approach introduced in Chapter 4.

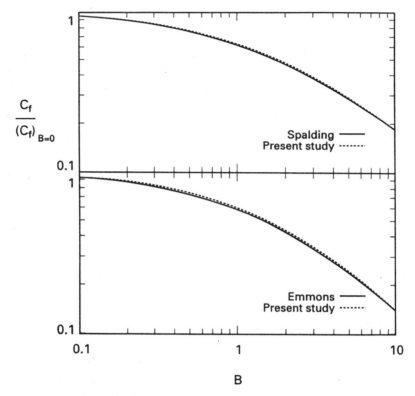

FIGURE 6.5: Spalding and Emmons versus present study.

6.1.2   Turbulent Flame

Following the usual practice, decompose the instantaneous velocity, $\tilde{u}_i$, and the Schvab-Zeldovich property of a turbulent diffusion flame, $\tilde{b}$, into a temporal mean (denoted by capital letters) and fluctuations

$$\tilde{u}_i = U_i + u_i \text{ and } \tilde{b} = B + b .$$

For a homogeneous pure shear flow (in which all averages except U_i and B are independent of position and in which the mean strain rate S_{ij} is a constant), the mean kinetic energy of velocity fluctuations and the mean of the Schvab-Zeldovich property fluctuations yield

$$\mathcal{P} = \epsilon , \tag{6.44}$$

and

$$\mathcal{P}_\beta = \epsilon_\beta , \tag{6.45}$$

where

$$\mathcal{P} = -\overline{u_i u_j} S_{ij} \tag{6.46}$$

is the inertial production,

$$\epsilon = 2\nu_\beta \overline{s_{ij} s_{ij}} \tag{6.47}$$

is the dissipation of turbulent energy,

$$\mathcal{P}_\beta = -\overline{u_i b}\frac{\partial B}{\partial x_i} \tag{6.48}$$

and

$$\epsilon_\beta = D_\beta \overline{\left(\frac{\partial b}{\partial x_i}\right)\left(\frac{\partial b}{\partial x_i}\right)} \tag{6.49}$$

are the *production and dissipation of the Schvab-Zeldovich property*, respectively. Note that the incorporation of boundary mass transfer into Eqs. (6.44) and (6.45) are taken into account in terms of ν_β and D_β.

On dimensional grounds, Eqs. (6.44) and (6.45) lead to

$$\mathcal{P} \sim \frac{u^3}{\ell} \sim \nu_\beta \frac{u^2}{\lambda^2} \sim \epsilon \tag{6.50}$$

and

$$u_\beta \frac{b^2}{\ell} \sim D_\beta \frac{b^2}{\lambda_\beta^2} . \tag{6.51}$$

where λ and λ_β are Taylor microscales associated with momentum and the Schvab-Zeldovich property.

From Eq. (6.50),

$$\frac{\lambda}{\ell} \sim Re_\beta^{-1/2} . \tag{6.52}$$

Also, from Eq. (6.50), an inertial estimate for u is

$$u \sim (\epsilon\ell)^{1/3} , \tag{6.53}$$

and, a viscous estimate is

$$u \sim \lambda(\epsilon/\nu_\beta)^{1/2} .$$

(6.54)

Elimination of velocity between Eqs. (6.53) and (6.54) yields

$$\lambda \sim \ell^{1/3} \left(\frac{\nu_\beta^3}{\epsilon} \right)^{1/6} .$$

(6.55)

For the isotropic flow, λ and ℓ replaced with one scale,

$$\left(\frac{\lambda}{\ell} \right) \to \eta ,$$

(6.56)

leads to a Kolmogorov scale ,

$$\eta \sim \left(\frac{\nu_\beta^3}{\epsilon} \right)^{1/4} .$$

(6.57)

Elimination of ν_β^3/ϵ between Eqs. (6.55) and (6.57) yields

$$\frac{\eta}{\ell} \sim \left(\frac{\lambda}{\ell} \right)^{3/2} ,$$

(6.58)

or, in terms of Eq. (6.52),

$$\frac{\eta}{\ell} \sim Re_\beta^{-3/4} .$$

(6.59)

Now, assume the momentum in terms of ν_β and the b-property in terms of D_β be similar in a manner similar to the momentum and thermal energy of forced convection. Then, in terms of

$$u_\beta = u \left(\frac{\lambda_\beta}{\lambda} \right) ,$$

(6.60)

Eq. (6.51) becomes

$$\frac{u}{\ell} \left(\frac{\lambda_\beta}{\lambda} \right) \sim \frac{D_\beta}{\lambda_\beta^2}$$

(6.61)

which leads, in terms of Eq. (6.54), to

$$\frac{\lambda_\beta}{\ell} \left(\frac{\epsilon}{\nu_\beta} \right)^{1/2} \sim \frac{D_\beta}{\lambda_\beta^2} ,$$

(6.62)

or, after some rearrangement, to a flame Taylor scale,

$$\lambda_\beta \sim \ell^{1/3} \left(\frac{\nu_\beta D_\beta^2}{\epsilon} \right)^{1/6} . \tag{6.63}$$

The ratio of Eqs. (6.55) and (6.63) gives

$$\frac{\lambda}{\lambda_\beta} = \left(\frac{\nu_\beta}{D_\beta} \right)^{1/3} = \sigma_\beta^{1/3} . \tag{6.64}$$

In the isotropic limit,

$$\left(\frac{\lambda_\beta}{\ell} \right) \to \eta_B , \tag{6.65}$$

and Eq. (6.63) is reduced to a flame Batchelor scale,

$$\eta_B \sim \left(\frac{\nu_\beta D_\beta^2}{\epsilon} \right)^{1/4} . \tag{6.66}$$

However, Eq. (6.56) already sets forth

$$\ell \to \eta . \tag{6.67}$$

In addition, let now

$$\lambda_\beta \to \eta_\beta . \tag{6.68}$$

Then, Eq. (6.63) is reduced, in terms of Eqs. (6.67) and (6.68), to another microscale,

$$\eta_\beta \sim \eta^{1/3} \left(\frac{\nu_\beta D_\beta^2}{\epsilon} \right)^{1/6} , \tag{6.69}$$

which, using Eq. (6.66), can be rearranged as

$$\eta_\beta \sim (\eta \eta_B^2)^{1/3} , \tag{6.70}$$

a mesomicroscale based on a weighted average of the Kolmogorov and Batchelor scales. Also, in terms of Eqs. (6.57) and (6.66),

$$\eta_\beta \sim \left(\frac{\nu_\beta^{5/3} D_\beta^{4/3}}{\epsilon} \right)^{1/4} . \tag{6.71}$$

Note that η_B is the isotropic scale of the isotropic b-balance while η_β is the isotropic scale of the lower limit of the homogeneous b-balance. The ratio of Eqs. (6.57) and (6.71), in a manner identical to that of Eqs. (6.55) and (6.63), yields (recall Eq. 6.64)

$$\frac{\eta}{\eta_\beta} = \sigma_\beta^{1/3} . \tag{6.72}$$

Experimental literature supports the relevance of η_β rather than η_B for mass transfer.

Now, let the turbulent forced diffusion flame over horizontal fuel be controlled by a sublayer of thickness η_β. Then, the averaged burning rate over a length ℓ of the fuel becomes

$$\frac{M'}{\rho D} \sim B \frac{\ell}{\eta_\beta} = C B \left(\frac{\ell}{\eta} \right) \left(\frac{\eta}{\eta_\beta} \right) , \tag{6.73}$$

or, in terms of Eqs. (6.35), (6.59) and (6.72),

$$\frac{M'}{\rho D} = \frac{C_1 B Re_\beta^{3/4} \sigma_\beta^{1/3}}{(1 + C_0 B)^n} , \tag{6.74}$$

or, in terms of Eqs. (6.22) and (6.24),

$$\frac{M'}{\rho D} = \frac{C_1 B}{(1 + C_0 B)^n} \left(\frac{Re}{1 + B'} \right)^{3/4} \left(\frac{1 + B'}{1 + B} \right)^{1/3} \sigma^{1/3} \tag{6.75}$$

which leads to

$$\frac{M'}{\rho D} = \frac{C_1 B Re^{3/4} \sigma^{1/3}}{(1 + C_0 B)^n (1 + B')^{5/12} (1 + B)^{1/3}} . \tag{6.76}$$

For $\chi \sim 1$ and $\sigma \sim 1$, $B' \sim B$ and Eq. (6.76) reduces to

$$\frac{M'}{\mu Re^{3/4}} = \frac{C_1 B}{(1 + C_0 B)^n (1 + B)^{3/4}} . \tag{6.77}$$

For $B = 0$, the experimental literature gives

$$C_1 = 0.0396 , \quad Re \leq 10^5 .$$

For $B \to \infty$, noting unavailability of a Spalding flavored turbulent study, assume the fuel consumption approaches a constant (different than Emmons). This requires

$$n = \frac{1}{4} .$$

Then,

$$\frac{M'}{\mu Re^{3/4}} = \frac{0.0396B}{(1 + C_0 B)^{1/4}(1 + B)^{3/4}} , \qquad (6.78)$$

and the skin friction,

$$C_f Re^{1/4} = \frac{0.0791}{(1 + C_0 B)^{1/4}(1 + B)^{3/4}} . \qquad (6.79)$$

Arpacı (1990a) on natural convection usually leads to $C_0 \ll 1$. The experimental literature on the present problem is quite scarce for a reliable estimate of C_0. Here, the past experiments on skin friction employed by Marxman (1967) are utilized for an estimate. A least squares fit leads also to a very small C_0 which is well within the uncertainty of the data. Accordingly,

$$C_0 \cong 0 \qquad (6.80)$$

is assumed. Then, to a first order,

$$\frac{C_f}{(C_f)_{B=0}} = \frac{1}{(1 + B)^{3/4}} , \qquad (6.81)$$

Fig. 6.6 shows the correlation of the data with Eq. (6.81). Extensive new data is needed before a reasonable estimate can be made for C_0.

6.2 BUOYANCY-DRIVEN FLOWS

First, consider a dimensional review of laminar flames which proves convenient in Section 6.2.2 on turbulent flames.

6.2.1 Laminar Flame

Reconsider the second part of Spalding (1954) on buoyancy-driven flames. The balance of momentum integrated over boundary layer thickness δ is

$$\frac{d}{dx} \int_0^\delta \rho u^2 \, dy + \left(\mu \frac{\partial u}{\partial y} \right)_w = g \int_0^\delta (\rho_\infty - \rho) \, dy , \qquad (6.82)$$

where ρ is the density, u the longitudinal velocity, μ the dynamic viscosity, and subscripts w and ∞ denote wall (fuel surface) and ambient

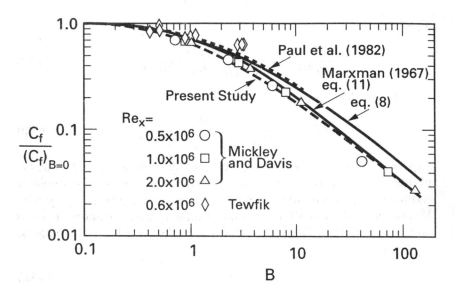

FIGURE 6.6: Correlation of skin friction with Eq. (6.81).

conditions. Also, the balance of the Schvab-Zeldovich (heat+oxidizer) property integrated over the boundary layer thickness δ_β is

$$\frac{d}{dx}\int_0^{\delta_\beta} \rho u(b_\infty - b)dy - B(\rho v)_w = \left(\rho D \frac{\partial b}{\partial y}\right)_w, \qquad (6.2)$$

where b and B being already defined with Eqs. (6.3) through (6.5).

On dimensional grounds, Eq. (6.82) yields

$$U\frac{U}{\ell}\delta + \nu\frac{U}{\delta} \sim g\left(\frac{\Delta\rho}{\rho}\right)\delta, \qquad (6.83)$$

U being a characteristic longitudinal velocity, and ℓ a length scale characterizing the direction of flow. Similarly, Eq. (6.2) yields for $U_\beta = U$,

$$U\frac{B}{\ell}\delta_\beta - v_w B \sim D\frac{B}{\delta_\beta}. \qquad (6.11)$$

In terms of the surface mass balance,

$$\rho v_w \sim \rho D \frac{B}{\delta_\beta}, \qquad (6.12)$$

Eq. (6.11) may be rearranged as

$$U\frac{B}{\ell} \sim D(1+B)\frac{B}{\delta_\beta^2} , \qquad (6.84)$$

and, in terms of the Squire postulate (recall Eq. 4.20) for buoyancy-driven flows,

$$\delta \sim \delta_\beta , \qquad (6.85)$$

Eq. (6.83) becomes

$$U\frac{U}{\ell} + \nu\frac{U}{\delta_\beta^2} \sim g\left(\frac{\Delta\rho}{\rho}\right) . \qquad (6.86)$$

This postulate has been well-tested in natural convection even for δ/δ_β differing considerably from unity. Its validity for the present problem will be justified later (in the discussion of Fig. 6.8). Also, because of the same b-gradient involved with Eqs. (6.11) and (6.12), the factor $(1+B)$ is independent of the dimensional arguments leading to Eq. (6.84). For notational convenience, let

$$D_\beta = D(1+B) . \qquad (6.16)$$

Then, Eq. (6.84) is reduced to

$$U\frac{B}{\ell} \sim D_\beta\frac{B}{\delta_\beta^2} . \qquad (6.87)$$

Clearly, Eqs. (6.86) and (6.87) can be directly obtained from the corresponding differential formulations, provided D_β is assumed for diffusivity in the latter.

A dimensionless number that describes buoyancy-driven diffusion flames may now be obtained by coupling Eqs. (6.86) and (6.87). Since velocity is a dependent variable for any buoyancy-driven flow, its elimination between these equations yields

$$\frac{\ell}{\delta_\beta^4}\left(1+\frac{D_\beta}{\nu}\right) \sim \frac{g}{\nu D_\beta}\left(\frac{\Delta\rho}{\rho}\right) \qquad (6.88)$$

which, in terms of a flame Schmidt number,

$$\sigma_\beta \sim \frac{\nu}{D_\beta} \qquad (6.89)$$

and a flame Rayleigh number,

$$Ra_\beta = \frac{g}{\nu D_\beta} \left(\frac{\delta\rho}{\rho}\right) \ell^3 \,, \tag{6.90}$$

may be rearranged as

$$\frac{\ell}{\delta_\beta} \sim \Pi_\beta^{1/4} \,, \tag{6.91}$$

where

$$\Pi_\beta \sim \left(\frac{\sigma_\beta}{1 + \sigma_\beta}\right) Ra_\beta \tag{6.92}$$

is a combined dimensionless number for diffusion flames. Actually, the numeral one in Eq. (6.92) is an unknown constant because of the dimensional nature of the foregoing arguments. Eq. (6.89) reflects this fact by its proportionality sign. Also, in view of

$$\frac{\Delta\rho}{\rho} \sim \frac{\rho_\infty - \rho_f}{\rho_f} = \frac{T_f - T_\infty}{T_\infty} \,, \tag{6.93}$$

the Rayleigh number may be more appropriately written as

$$Ra_\beta = \frac{g(T_f - T_\infty)\ell^3}{\nu D_\beta T_\infty} \,. \tag{6.94}$$

Now, in terms of a (fuel) mass transfer coefficient h_β,

$$Sh_\beta = \frac{h_\beta \ell}{D} \sim \frac{\ell}{\delta_\beta} \,, \tag{6.95}$$

Sh_β being a flame Sherwood number. Then, the fuel consumption in a laminar diffusion flame of size ℓ,

$$\frac{m'}{\rho D} = \frac{m''_w \ell}{\rho D} = Sh_\beta B \sim B\frac{\ell}{\delta_\beta} \,, \tag{6.96}$$

(m''_w being the fuel consumption per unit area) may be written in terms of Eq. (6.91) as

$$\frac{m'}{\rho D} \sim B\Pi_\beta^{1/4} \,, \tag{6.97}$$

or, explicitly,

$$\frac{m'}{\rho D} \sim B \left(\frac{\sigma_\beta}{1 + \sigma_\beta}\right)^{1/4} Ra_\beta^{1/4} , \qquad (6.98)$$

or, in terms of the usual Rayleigh number,

$$Ra = \frac{g}{\nu \alpha} \left(\frac{\Delta \rho}{\rho}\right) \ell^3 , \qquad (6.99)$$

as

$$\frac{m'}{\rho D Ra^{1/4}} \sim B \left(\frac{\sigma_\beta}{1 + \sigma_\beta}\right)^{1/4} \left(\frac{D}{D_\beta}\right)^{1/4} . \qquad (6.100)$$

Now, introduce the definition of the usual Schmidt number,

$$\sigma = \frac{\nu}{D} , \qquad (6.101)$$

and combine Eqs. (6.16) and (6.89) for

$$\frac{\sigma_\beta}{\sigma} \sim \frac{D}{D_\beta} = \frac{1}{1 + B} . \qquad (6.102)$$

Then, an equality replacing Eq. (6.100) may be written as

$$\frac{m'}{\rho D Ra^{1/4}} = \frac{C_1 B}{(C_0 + B)^{1/4}(1 + B)^{1/4}} , \qquad (6.103)$$

where C_0 and C_1 remain to be determined from a computer/laboratory experiment, or, from an analytical solution. However, before proceeding to these coefficients, a number of important facts can be deduced from Eq. (6.103).

For small values of B,

$$\lim_{B \to 0} \left(\frac{m'}{\rho D Ra^{1/4}}\right) \to B . \qquad (6.104)$$

For $B > 1$, inertial effects are negligible. Either eliminating the inertial term of the momentum balance (Eq. 6.86), or, noting Eq. (6.102) and the definition of

$$\sigma_\beta = \left(\frac{\text{Viscous force}}{\text{Inertial force}}\right) \left(\frac{\text{Flow of } B}{\text{Diffusion of } B}\right) , \qquad (6.105)$$

and letting $\sigma_\beta \to \infty$ in

$$\frac{\sigma_\beta}{1 + \sigma_\beta} = \frac{1}{1 + \sigma_\beta^{-1}} \to 1 , \qquad (6.106)$$

Eq. (6.103) is reduced to

$$\frac{m'}{\rho D Ra^{1/4}} \to B^{3/4} , \quad B > 1 , \qquad (6.107)$$

a well-known but so far assumed to be an experimentally supported empirical result. For $B \gg 1$,

$$\lim_{B \to \infty} \left(\frac{m'}{\rho D Ra^{1/4}} \right) \to B^{1/2} . \qquad (6.108)$$

Hereafter, Eq. (6.103) is called LM (Laminar Model).

Now, for numerical values of C_0 and C_1, reconsider Spalding (1954) which, after a minor algebraic correction,[1] yields for the fuel consumption averaged over a length ℓ,

$$\frac{m'}{\rho D Ra^{1/4}} = \frac{0.34115 B}{\left[\frac{1}{a^2} \left(\frac{1+B}{5.5-a} \right)^2 + 0.6\sigma \frac{1}{a^3} \left(\frac{1+B}{5.5-a} \right) \right]^{1/4}} , \qquad (6.109)$$

or, with some arrangement,

$$\frac{m'}{\rho D Ra^{1/4}} = \frac{0.34115 \left[a(5.5-a) \right]^{1/2} B}{\left[\left(1 + \frac{0.6(5.5-a)\sigma}{a} \right) + B \right]^{1/4} (1+B)^{1/4}} , \qquad (6.110)$$

where

$$a = \frac{2}{B} \left[\left(1 + \frac{3}{2}B \right)^{1/2} - 1 \right] . \qquad (6.111)$$

A comparison of Eqs. (6.103) and (6.110) readily suggests that C_0 and C_1 are not actually constants but depend on B, as to be expected in view of the B-dependence of the b-profiles (Fig. 6.7). Thus,

$$\frac{m'}{\rho D Ra^{1/4}} = \frac{C_1(B)B}{[C_0(B) + B]^{1/4} (1+B)^{1/4}} , \qquad (6.112)$$

---

[1] Each factor $(3.25 + a)$ in Spalding needs to be replaced by $(5.5 - a)$.

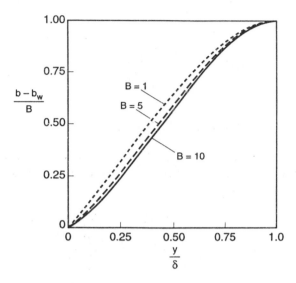

FIGURE 6.7: $(b - b_w)/B$-profile versus dimensionless boundary layer thickness as a function of B.

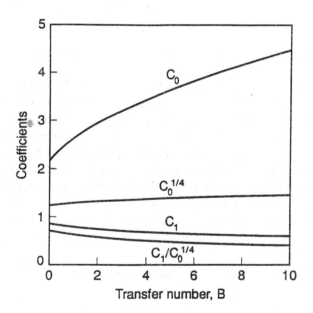

FIGURE 6.8: C_0, $(C_0)^{1/4}$, C_1, and $C_1/(C_0)^{1/4}$ versus B.

where

$$C_0(B) = 1 + \frac{0.6(5.5 - a)\sigma}{a} \qquad (6.113)$$

and

$$C_1(B) = 0.34115\,[a(5.5 - a)]^{1/2} \;. \qquad (6.114)$$

Figure 6.8 shows the dependence of C_0 and C_1 on B. However, when Spalding's study is repeated with a linear b-profile, Eqs. (6.109) and (6.110) are reduced to

$$\frac{m'}{\rho D Ra^{1/4}} = \frac{0.34115B}{\left[\left(\dfrac{1+B}{5.25}\right)^2 + 0.6\sigma\left(\dfrac{1+B}{5.25}\right)\right]^{1/4}} , \qquad (6.115)$$

and

$$\frac{m'}{\rho D Ra^{1/4}} = \frac{0.34115(5.25)^{1/2}B}{[(1 + (0.6)(5.25)\sigma) + B]^{1/4}(1 + B)^{1/4}} . \qquad (6.116)$$

and the B-dependence[2] of Eq. (6.116) turns out to be identical to that of Eq. (6.103).

For an evaluation of the constants involved with LM, first consider the practical range of B. Assuming an approximate upper bound of 10 for B, the fuel consumption versus B obtained from Spalding, as well as from the computational study of Kim, de Ris and Kroesser (1971) (say KRK), are plotted in Fig. 6.9. Since Spalding employs the Squire postulate but KRK does not, the close agreement indicates the validity of this postulate also for buoyancy-driven diffusion flames. An over-prediction of the burning rate by Spalding's approach should be expected in view of the fact that Spalding assumes a constant $\Delta\rho/\rho$ based on maximum buoyancy. Thus, it is reasonable to assume that KRK is numerically

---

[2]It is interesting to note that for small values of B, $a \to \dfrac{3}{2}$ and Eq. (6.110) is reduced to

$$\frac{m'}{\rho D Ra^{1/4}} = \frac{0.34115(6)^{1/2}B}{[(1 + (0.6)(2.6667)\sigma) + B]^{1/4}(1 + B)^{1/4}}$$

whose B-dependence is also identical to that of Eq. (6.103).

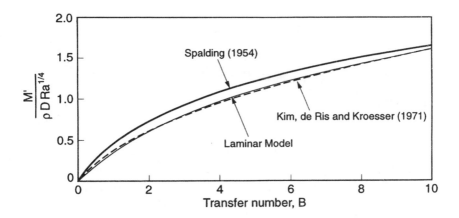

FIGURE 6.9: Comparison of a number of models for laminar fuel consumption versus B.

somewhat more accurate than Spalding because of its computational rather than approximate integral solution. Here, a least-squares fitting of Eq. (6.103) to KRK over the range $0 \leq B \leq 10$ should be expected. Instead, a simpler curve-fitting procedure in parallel to that to be employed for turbulent flame is followed here.

In the limit of $B \to 0$, fuel consumption approaches the heat transfer correlation, that is,

$$\lim_{B \to 0} \left(\frac{m'}{\rho D B} \right) \to Nu . \qquad (6.117)$$

The computational solution, available for the classical problem of natural convection next to a vertical flat plate gives, for $Pr = 0.73$,

$$Nu = 0.52 Ra^{1/4} \qquad (6.118)$$

(see, for example, Ede, 1967). Thus, in view of Eq. (6.103),

$$\lim_{B \to 0} \left(\frac{m'}{\rho D B} \right) \to \frac{C_1}{C_o^{1/4}} Ra^{1/4} , \qquad (6.119)$$

Eqs. (6.117) and (6.118) yield

$$\frac{C_1}{C_o^{1/4}} = 0.52 . \qquad (6.120)$$

Then, $1/C_0 = 0.90$ is found by matching Eq. (6.103) to KRK at $B = 2$. With these values, Eq. (6.103) becomes

$$\frac{m'}{\rho D Ra^{1/4}} = \frac{0.52B}{(1+0.90B)^{1/4}(1+B)^{1/4}} , \quad B \leq 10 . \quad (6.121)$$

Figure 6.9 shows the close agreement between Eq. (6.121) and KRK despite one point-matching. Also interesting is the fact that the model is obtained with some mean values of C_0 and C_1 without taking their dependence on B into account. However, since the dependence of C_1 on B is weak and that of C_0 is weakened by its fractional power involved with Eq. (6.121), the result is not surprising. This is indeed an important fact because, under turbulent conditions, the dependence of these coefficients on B is difficult to estimate.

The foregoing dimensional arguments on laminar diffusion flames are extended to buoyancy-driven turbulent diffusion flames and pool fires by following the recent developments on turbulent natural convection (Arpacı, 1986, 1990, 1995).

6.2.2 Turbulent Flame

Following the usual practice, decompose the instantaneous velocity and the Schvab-Zeldovich (heat+oxidizer) property of a buoyancy-driven, turbulent diffusion flame into a temporal mean (denoted by capital letters) and fluctuations

$$\tilde{u}_i = U_i + u_i \text{ and } \tilde{b} = B + b .$$

For a homogeneous pure shear flow (in which all averaged except U_i and B are independent of position and in which S_{ij} is a constant), the mean kinetic energy of velocity fluctuations and the root mean square of turbulent Schvab-Zeldovich property yield

$$\mathcal{B} = \mathcal{P} + \epsilon \quad (6.122)$$

and

$$\mathcal{P}_\beta = \epsilon_\beta , \quad (6.45)$$

where

$$\mathcal{B} = -g_i \overline{u_i \theta}/\Theta_0 \quad (6.123)$$

is the buoyant production (imposed),

$$\mathcal{P} = -\overline{u_i u_j} S_{ij} \tag{6.46}$$

is the inertial production (induced),

$$\epsilon = 2\nu \overline{s_{ij} s_{ij}} \tag{6.47}$$

is the dissipation of turbulent energy, and

$$\mathcal{P}_\beta = -\overline{u_i b} \frac{\partial B}{\partial x_i} \tag{6.48}$$

and

$$\epsilon_\beta = D_\beta \overline{\left(\frac{\partial b}{\partial x_i}\right)\left(\frac{\partial b}{\partial x_i}\right)} \tag{6.49}$$

are the production and dissipation of the Schvab-Zeldovich property, respectively. Note that the incorporation of boundary mass transfer into the b-balance is taken into account by considering a b-dissipation in terms of D_β. For buoyancy-driven flows, kinetic dissipation retains its usual form.

On dimensional grounds, Eqs. (6.122) and (6.45) lead to

$$\mathcal{B} \sim \frac{u^3}{\ell} + \nu \frac{u^2}{\lambda^2} , \tag{6.124}$$

and

$$u_\beta \frac{b^2}{\ell} \sim D_\beta \frac{b^2}{\lambda_\beta^2} . \tag{6.51}$$

where λ and λ_β are Taylor microscales associated with momentum and the Schvab-Zeldovich property.

Now, for a buoyancy-driven turbulent diffusion flame, following the Squire postulate, assume

$$u \sim u_\beta , \quad \lambda \sim \lambda_\beta . \tag{6.125}$$

Then, elimination of velocity between Eqs. (6.124) and (6.51), gives

$$\lambda_\beta \sim \ell^{1/3} \left(1 + \sigma_\beta\right)^{1/6} \left(\frac{D_\beta^3}{\mathcal{B}}\right)^{1/6} . \tag{6.126}$$

Under conditions of isotropic flow,

$$\left(\frac{\lambda_\beta}{\ell} \right) \to \eta_\beta , \qquad (6.127)$$

and, Eq. (6.126) is reduced to a flame Kolmogorov scale,

$$\eta_\beta \sim (1 + \sigma_\beta)^{1/4} \left(\frac{D_\beta^3}{\mathcal{B}} \right)^{1/4} , \qquad (6.128)$$

where, on dimensional grounds,

$$\mathcal{B} \sim gu\theta/\Theta_0 , \qquad (6.129)$$

Θ_0 being the temperature of isobaric ambient. For $\sigma_\beta \to 0$, Eq. (6.128) is reduced in form to an Oboukhov-Corrsin scale,

$$\eta_\beta \sim \left(\frac{D_\beta^3}{\mathcal{B}} \right)^{1/4} . \qquad (6.130)$$

Also, for $\sigma_\beta \to \infty$, Eq. (6.128) is reduced in form to a flame Batchelor scale,

$$\eta_\beta \sim \left(\frac{\sigma_\beta D_\beta^3}{\mathcal{B}} \right)^{1/4} = \left(\frac{\nu D_\beta^2}{\mathcal{B}} \right)^{1/4} . \qquad (6.131)$$

To proceed further, let

$$\theta \sim \Delta T , \qquad (6.132)$$

ΔT being the imposed temperature difference, and note, for gaseous media,

$$\Theta_0^{-1} = \beta . \qquad (6.133)$$

Then, Eq. (6.129) becomes

$$\mathcal{B} \sim g\beta u \Delta T , \qquad (6.134)$$

or, in view of Eq. (6.51) for $u_\beta \sim u$,

$$\mathcal{B} \sim g\beta D_\beta \ell \Delta T/\lambda_\beta^2 . \qquad (6.135)$$

Insertion of Eq. (6.135) into Eq. (6.126) leads to the Taylor scale depending on the buoyant force rather than buoyant energy,

$$\lambda_\beta \sim \ell^{1/4} (1 + \sigma_\beta)^{1/4} \left(\frac{D_\beta^2}{g\beta\Delta T} \right)^{1/4} , \tag{6.136}$$

or, under the isotropy stated by Eq. (6.127), to a Kolmogorov scale,

$$\eta_\beta \sim (1 + \sigma_\beta)^{1/3} \left(\frac{D_\beta^2}{g\beta\Delta T} \right)^{1/3} . \tag{6.137}$$

Now, the Taylor and Kolmogorov scales for any σ_β may be rearranged in terms of Π_β as

$$\frac{\lambda_\beta}{\ell} \sim \Pi_\beta^{-1/4} \tag{6.138}$$

and

$$\frac{\eta_\beta}{\ell} \sim \Pi_\beta^{-1/3} . \tag{6.139}$$

Let the turbulent diffusion flame near a vertical fuel or the pool fire over a horizontal fuel be controlled by a turbulent sublayer. Assume the thickness of this layer be characterized by η_β. Then, in terms of Eq. (6.25), the averaged fuel consumption is found to be

$$\frac{m'}{\rho D} \sim B\frac{\ell}{\eta_\beta} \sim B\Pi_\beta^{1/3} , \tag{6.140}$$

or, explicitly,

$$\frac{m'}{\rho D} \sim B \left(\frac{\sigma_\beta}{1 + \sigma_\beta} \right)^{1/3} Ra_\beta^{1/3} , \tag{6.141}$$

or, in terms of the usual Rayleigh number,

$$\frac{m'}{\rho D Ra^{1/3}} \sim B \left(\frac{\sigma_\beta}{1 + \sigma_\beta} \right)^{1/3} \left(\frac{D}{D_\beta} \right)^{1/3} . \tag{6.142}$$

Now, rearranging Eq. (6.142) in terms of Eq. (6.16),

$$\frac{m'}{\rho D Ra^{1/3}} = \frac{C_1 B}{(C_0 + B)^{1/3}(1 + B)^{1/3}} , \tag{6.143}$$

where C_0 and C_1 (different than those of Eq. 6.103) are to be determined from experimental literature. The 1/3-power law of Rayleigh number in pool fires is supported experimentally (Kanury, 1975, Lockwood and Corlett, 1987, Alpert, 1977). Hereafter Eq. (6.143) is called TM (Turbulent Model).

Now, in a manner similar to laminar flames, the three distinct regimes of turbulent flames may be identified. For small values of B,

$$\lim_{B \to 0} \left(\frac{m'}{\rho D Ra^{1/3}} \right) \to B . \tag{6.144}$$

For $B > 1$, inertial effects are negligible and Eq. (6.143) is reduced to

$$\frac{m'}{\rho D Ra^{1/3}} \to B^{2/3} . \tag{6.145}$$

For $B \gg 1$,

$$\lim_{B \to \infty} \left(\frac{m'}{\rho D Ra^{1/3}} \right) \to B^{1/3} . \tag{6.146}$$

The experimental data on small fires (see Corlett, 1968, 1970, de Ris and Orloff, 1972, Burgess et al., 1961) appears to correlate well with TM as shown in Fig. 6.10. The original figure is taken from de Ris and Orloff who rearranged Fig. 5 of Corlett (1970) for ethane-nitrogen flames burning above a 10.16 cm diameter burner and compared with their model. The open symbols in the original figure for pure ethane are deleted here since they include a radiative heat transfer component towards the burner surface. Remaining data from Corlett represents the dominant convective component of the surface heat transfer. Half-filled symbols indicate increasing heat transfer with increasing velocity of gases leaving the burner surface. Also included in Fig. 6.10 are two data points from Burgess et al. for liquid methanol and liquid butane as shown by open symbols. The low B-range is in the vicinity of extinction of the flame.

Arpacı and Dec (1987) has recently demonstrated, with a correlation on natural convection, the sensitivity of C_0 to experimental data. A preliminary attempt for the evaluation of C_0 and C_1 by a least-square fitting of Eq. (6.143) to Corlett's data demonstrates a similar sensitivity. Here, following the approach taken in the preceding section on laminar flames, the value of $C_1/C_o^{1/3} = 0.16$ is taken from the recent data of

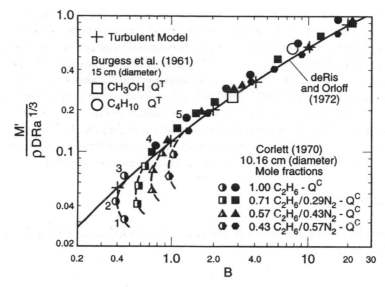

FIGURE 6.10: A correlation of the experimental data with Eq. (6.147).

Fujii and Imura (1972). Then, at $B = 5$, $1/C_0 = 0.05$ is evaluated by fitting Eq. (6.143) to Corlett's data. With these values, Eq. (6.143) becomes

$$\frac{m'}{\rho D Ra^{1/3}} = \frac{0.16B}{(1 + 0.05B)^{1/3}(1 + B)^{1/3}} \tag{6.147}$$

which agrees well with correlation

$$\frac{m'}{\rho D Ra^{1/3}} = 0.16B \left[\frac{\ell n (1 + B)}{B} \right]^{2/3} \tag{6.148}$$

given by de Ris and Orloff (1972)on the basis of the stagnant film theory coupled with empirically assumed 2/3-power law. The maximum difference between two correlations remains less than 1.8% for the entire B-range. The agreement, despite the fact that they are developed by following quite different arguments, is remarkable. Both models predict only the region beyond bifurcation for $B \geq 1$ (Arpacı and Selamet, 1991).

So far, the proposed models for laminar and turbulent flames and fires exclude any effect of radiation. Because of different intrinsic nature of radiation and conduction (or any diffusion), the Schvab-Zeldovich transformation used in the present study no longer applies to radiation-affected flames. On intuitive grounds, the emission effect of radiation has been already incorporated into the heat of combustion and the latent heat of evaporation by fractional lowering (say γ and ψ) of these properties (see Kanury, 1975). Following Arpacı and coworkers (see, for example, Arpacı, 1991 and the references cited therein), and Selamet and Arpacı (1989), the optical thickness and scattering can be incorporated into γ and ψ. However, because of lack of experimental data, no attempt is made here to demonstrate their influence on γ and ψ.

PROBLEMS

6-1) Develop the microscales of buoyancy-driven diffusion flames for a specified mass flux, m''_w.

CHAPTER 7

UNSTEADY TURBULENCE

Increased experimental and theoretical attention is being paid to the effect of organized waves on turbulent shear flows. These waves may be externally imposed, say, by a vibrating ribbon near walls of a channel (Hussain and Reynolds 1970, 1972, hereafter called HR, HR2) or may be internally induced, say, in a pulse combustor (Dec and Keller, 1989,1990). Also, the dynamical equations which govern oscillating (with single frequency on the mean) turbulent shear flows have been developed (Reynolds and Hussain, 1972, hereafter called RH). As expected, these equations require additional information about the wave-induced fluctuations in Reynolds stresses before a closed system can be obtained. All closure attempts appear to be statistical, empirical or the usual normal-mode approach of instability theories. Thermal microscales for these flows and their significance for heat transfer in pulsating turbulent flows so far appear to be left untreated except for the recent work of Arpacı, et al. (1993) and are the motivations of this chapter.

7.1 DYNAMICS OF UNSTEADY TURBULENCE

Our starting point is the dynamical equations for unsteady turbulent shear flows which are already available in the literature (see HR and RH). A brief review of the development leading to these equations is borrowed from the literature and is given below.

Let an instantaneous quantity of turbulence f($\boldsymbol{x}$,t) be decomposed as (Fig. 7.1)

$$f(\boldsymbol{x},t) = \overline{f}(\boldsymbol{x}) + \tilde{f}(\boldsymbol{x},t) + f'(\boldsymbol{x},t) \tag{7.1}$$

143

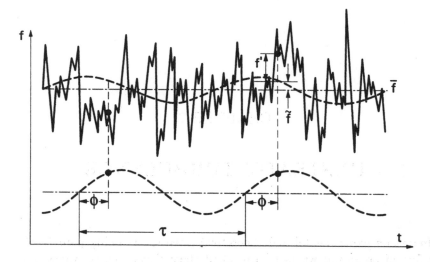

FIGURE 7.1: Decomposition of unsteady turbulence. From Hussain and Reynolds (1970).

where $\overline{f}$ is the mean (temporal average), $\tilde{f}$ the (statistical) contribution of the organized wave, and f' the turbulent (random) fluctuations. The *temporal average* is

$$\overline{f}(\boldsymbol{x}) = \lim_{T \to \infty} \frac{1}{T} \int_0^T f(\boldsymbol{x}, t) dt , \qquad (7.2)$$

the *phase average* is

$$\langle f(\boldsymbol{x}, t) \rangle = \lim_{N \to \infty} \frac{1}{N} \sum_{n=0}^{N} f(\boldsymbol{x}, t + n\tau) , \qquad (7.3)$$

where τ is the (specified) period of the imposed or induced wave. The phase average is the average over a large ensemble of points having the same phase with respect to the specified wave. Then

$$\tilde{f} = \langle f \rangle - \overline{f} , \qquad (7.4)$$

that is, the phase-averaging (ignores the background turbulence and) filters only the coherent oscillations from the instantaneous turbulent motion. The following properties resulting from time and phase averaging,

$$\langle f' \rangle = 0 , \quad \overline{\tilde{f}} = 0 , \quad \overline{f'} = 0 ,$$

$$\overline{f g} = \overline{f}\,\overline{g}\,,\quad \langle \tilde{f} g \rangle = \tilde{f}\langle g \rangle\,,\quad \langle \overline{f} g \rangle = \overline{f}\langle g \rangle\,, \tag{7.5}$$

$$\overline{\langle f \rangle} = \overline{f}\,,\quad \langle \overline{f} \rangle = \overline{f}\,,\quad \overline{\tilde{f} g'} = \langle \tilde{f} g' \rangle = 0\,,$$

are needed in the development of the dynamic equations.

The continuity, momentum and kinetic energy satisfied by U_i, $\tilde{u}_i$ and u_i', and the stability of $\tilde{u}_i$ to infinitesimal disturbances are discussed in RH. [1] We need the phase averaged kinetic energy associated with u_i' which satisfies

$$\frac{DK}{Dt} = \frac{\partial \mathcal{D}_j}{\partial x_j} + \mathcal{P} - \epsilon \tag{7.6}$$

where

$$\mathcal{D}_j = \left\langle u_j' \left(\frac{p'}{\rho} + \frac{1}{2} u_i' u_i' \right) \right\rangle + \langle \tilde{u}_j \frac{1}{2} u_i' u_i' \rangle - 2\nu \langle u_i' s_{ij}' \rangle \tag{7.7}$$

denotes the diffusion,

$$\mathcal{P} = -\langle u_i' u_j' \rangle \left(\overline{S}_{ij} + \tilde{S}_{ij} \right) \tag{7.8}$$

the production, and

$$\epsilon = 2\nu \langle s_{ij}' s_{ij}' \rangle \tag{7.9}$$

the dissipation of turbulent energy. For a homogeneous, pure shear flow, Eq. (7.6) reduces to

$$\frac{\partial K}{\partial t} = \mathcal{P} - \epsilon\,. \tag{7.10}$$

7.2 KINETIC SCALES

Let the phase averaged rms value of u_i' be u, the amplitude and frequency of $\tilde{u}_i$ be U_0, and ω, respectively. On dimensional grounds, rearrange Eq. (7.10) in terms of global[2] scales as

$$\omega u^2 + u^2 \left(\frac{U + U_0}{\ell} \right) \sim \nu \frac{u^2}{\lambda^2} \sim \epsilon\,, \tag{7.11}$$

---

[1]Following the notation of the preceding chapters, U_i, rather than $\overline{u}_i$, is used for the mean velocity.

[2]Volume-averaged.

where ℓ, and λ respectively denote an integral scale and the Taylor scale. On dimensional grounds, further assume

$$u \sim U + U_0 \tag{7.12}$$

by which Eq. (7.12) becomes

$$\omega u^2 + \frac{u^3}{\ell} \sim \nu \frac{u^2}{\lambda^2} \sim \epsilon , \tag{7.13}$$

where the first term denotes the temporal and the second spatial production. Eq. (7.13), rearranged relative to spatial production, gives

$$\frac{u^3}{\ell} \left(1 + \frac{\omega \ell}{u} \right) \sim \nu \frac{u^2}{\lambda^2} \sim \epsilon . \tag{7.14}$$

Inertial and viscous estimates for velocity obtained from Eq. (7.14),

$$u \sim \frac{(\epsilon \ell)^{1/3}}{\left(1 + \frac{\omega \ell}{u} \right)^{1/3}} \sim \lambda \left(\frac{\epsilon}{\nu} \right)^{1/2} \tag{7.15}$$

lead to a Taylor scale,

$$\lambda \sim \frac{\ell^{1/3} \left(\dfrac{\nu^3}{\epsilon} \right)^{1/6}}{\left(1 + \dfrac{\omega \ell}{u} \right)^{1/3}} \tag{7.16}$$

whose isotropic limit for

$$\left(\frac{\lambda}{\ell} \right) \to \eta , \quad u \to v , \tag{7.17}$$

leads, in view of

$$\tau^{-1} = \frac{\eta}{v} = \frac{\lambda}{u} \sim \left(\frac{\nu}{\epsilon} \right)^{1/2} , \tag{7.18}$$

to a Kolmogorov scale,

$$\eta \sim \frac{\left(\dfrac{\nu^3}{\epsilon} \right)^{1/4}}{\left[1 + \omega \left(\dfrac{\nu}{\epsilon} \right)^{1/2} \right]^{1/2}} . \tag{7.19}$$

Two limits of this scale,

$$\lim_{\omega \to 0} \eta \to \left(\frac{\nu^3}{\epsilon}\right)^{1/4} \tag{7.20}$$

$$\lim_{\omega \to \infty} \eta \to \left(\frac{\nu}{\omega}\right)^{1/2} \tag{7.21}$$

are the usual Kolmogorov and Stokes scales. Note that the latter, being independent of the effect of a mean velocity, exists under any (laminar or turbulent) flow condition.

Also, eliminating ν^3/ϵ between Eqs. (7.16) and (7.19), we have

$$\left(\frac{\eta}{\lambda}\right)^2 \sim \frac{\lambda}{\ell} \left(\frac{1 + \dfrac{\omega\ell}{u}}{1 + \dfrac{\omega\lambda}{u}}\right) \tag{7.22}$$

whose limit for $\omega \to 0$,

$$\left(\frac{\eta}{\lambda}\right)^2 \sim \frac{\lambda}{\ell} , \tag{7.23}$$

is used by Tennekes (1968) to construct a model for intermittency in steady (on the mean) turbulent flows. In terms of Eq. (7.22), the Tennekes model can be extended to unsteady flows. The next section deals with the thermal dynamics of unsteady turbulence which so far appears to be left untreated except for the recent work of Arpacı et al. (1996).

7.3 THERMAL DYNAMICS AND SCALES

In a manner similar to velocity, let the instantaneous temperature be split as

$$\theta = \bar{\theta} + \tilde{\theta} + \theta' , \tag{7.24}$$

$\bar{\theta}$ denoting the temporal average, $\tilde{\theta}$ the organized wave and θ' the turbulent (random) fluctuations. Here, we are interested in the behavior of θ', and for a positive measure for this behavior, consider

$$\mathcal{K}_\theta = \frac{1}{2}\langle\theta'\theta'\rangle \tag{7.25}$$

which satisfies

$$\frac{DK_\theta}{Dt} = -\frac{\partial(\mathcal{D}_\theta)_j}{\partial x_j} + \mathcal{P}_\theta - \epsilon_\theta , \tag{7.26}$$

where

$$(\mathcal{D}_\theta)_j = -\alpha\frac{\partial}{\partial x_j}\langle\frac{1}{2}\theta'\theta'\rangle + \langle\tilde{u}_j\frac{1}{2}\theta'\theta'\rangle + \langle u_j'\frac{1}{2}\theta'\theta'\rangle \tag{7.27}$$

denotes the thermal flux,

$$\mathcal{P}_\theta = -\langle u_j'\theta'\rangle\left(\frac{\partial\overline{\theta}}{\partial x_j} + \frac{\partial\tilde{\theta}}{\partial x_j}\right) , \tag{7.28}$$

the thermal production, and

$$\epsilon_\theta = \alpha\left\langle\left(\frac{\partial\theta'}{\partial x_j}\right)\left(\frac{\partial\theta'}{\partial x_j}\right)\right\rangle \tag{7.29}$$

the thermal dissipation. The details of the rather lengthy development leading to Eq. (7.26) are left to the reader (see Prob. 7.1). For homogeneous flows, Eq. (7.26) is reduced to

$$\frac{\partial K_\theta}{\partial t} = \mathcal{P}_\theta - \epsilon_\theta . \tag{7.30}$$

Now, we proceed to the thermal microscales associated with Eq. (7.30). Let the rms value of θ' be θ, the amplitude and frequency of $\tilde{\theta}$ be Θ_0 and ω, respectively. On dimensional grounds, we get from Eq. (7.30), for global scales,

$$\omega\theta^2 + u_\theta\theta\left(\frac{\Theta + \Theta_0}{\ell}\right) \sim \alpha\frac{\theta^2}{\lambda_\theta^2} , \tag{7.31}$$

λ_θ being a thermal Taylor scale. Assuming further

$$\theta \sim \Theta + \Theta_0 , \tag{7.32}$$

Eq. (7.31) is reduced to

$$\omega\theta^2 + u_\theta\frac{\theta^2}{\ell} \sim \alpha\frac{\theta^2}{\lambda_\theta^2} , \tag{7.33}$$

u_θ being a characteristic velocity for enthalpy flow which is different than that for momentum flow. For a spatially dominated thermal production, we may rearrange Eq. (7.33) as

$$u_\theta \frac{\theta^2}{\ell} \left(1 + \frac{\omega\ell}{u_\theta} \right) \sim \alpha \frac{\theta^2}{\lambda_\theta^2} . \tag{7.34}$$

Now, the ratio between Eqs. (7.14) and (7.34) gives

$$\frac{u}{u_\theta} \left(\frac{1 + \dfrac{\omega\ell}{u}}{1 + \dfrac{\omega\ell}{u_\theta}} \right) \sim \frac{\nu}{\alpha} \left(\frac{\lambda_\theta}{\lambda} \right)^2 . \tag{7.35}$$

The development that follows applies to fluids with $Pr \geq 1$. For $\omega \to 0$, the analogy between momentum and heat is well-known and leads to

$$\frac{u}{u_\theta} \sim \frac{\lambda}{\lambda_\theta} \sim Pr^{1/3} , \quad Pr \geq 1 . \tag{7.36}$$

Then, Eq. (7.35) becomes, in terms of Eq. (7.36),

$$\frac{\lambda}{\lambda_\theta} = Pr^{1/3} \left(\frac{1 + \dfrac{\omega\ell}{u_\theta}}{1 + \dfrac{\omega\ell}{u}} \right)^{1/3} . \tag{7.37}$$

Also, it can be shown, following arguments similar to those leading to the mesomicroscale introduced in Chapter 4, that

$$\frac{\eta}{\eta_\theta} = Pr^{1/3} \left(\frac{1 + \dfrac{\omega\lambda}{u_\theta}}{1 + \dfrac{\omega\lambda}{u}} \right)^{1/3} . \tag{7.38}$$

or, in terms of Eq. (7.36),

$$\frac{\eta}{\eta_\theta} = Pr^{1/3} \left[1 + \frac{\omega\lambda}{u} \left(Pr^{1/3} - 1 \right) \right]^{1/3} , \quad Pr \geq 1 . \tag{7.39}$$

Furthermore, for local scales, Eq. (7.34) becomes

$$u_\theta \frac{\theta^2}{\lambda} \left(1 + \frac{\omega\lambda}{u_\theta} \right) \sim \alpha \frac{\theta^2}{\eta_\theta^2} , \tag{7.40}$$

and the ratio between Eqs. (7.34) and (7.40) gives

$$\left(\frac{\eta_\theta}{\lambda_\theta}\right)^2 \sim \frac{\lambda}{\ell}\left(\frac{1 + \dfrac{\omega\ell}{u_\theta}}{1 + \dfrac{\omega\lambda}{u_\theta}}\right) \tag{7.41}$$

whose limit for $\omega \to 0$,

$$\left(\frac{\eta_\theta}{\lambda_\theta}\right)^2 \sim \frac{\lambda}{\ell} \tag{7.42}$$

extends the Tennekes model to thermal intermittency discussed in Chapter 4 (recalling Eqs. 4.119 and 4.120).

The foregoing kinetic and thermal scales are employed in the next section dealing with models for skin friction and heat transfer in oscillating turbulent flows.

7.4   SKIN FRICTION AND HEAT TRANSFER

Let the momentum sublayer of a turbulent flow be characterized by η, and the entire dissipation be confined to this layer. Then, the usual definition of skin friction, written in terms of local scales,

$$f = \frac{\tau_w}{\dfrac{1}{2}\rho u^2} \sim \frac{\mu\dfrac{u}{\eta}}{\dfrac{1}{2}\rho u^2} \tag{7.43}$$

or,

$$\frac{1}{2}f \sim \left(\frac{\ell}{\eta}\right) Re^{-1} \tag{7.44}$$

where

$$Re = \frac{u\ell}{\nu} \tag{7.45}$$

is the Reynolds number based on u and ℓ. Now, rearrange Eq. (7.14) and (7.23) as

$$\frac{\eta}{\ell} \sim \left(\frac{\lambda}{\ell}\right)^{3/2} \tag{7.46}$$

and

$$\frac{\ell}{\lambda} \sim Re^{1/2} \left(1 + \frac{\omega \ell}{u}\right)^{1/2} . \qquad (7.47)$$

Then, for small ω, Eq. (7.44) becomes, in terms of Eqs. (7.46) and (7.47),

$$\frac{1}{2}f \sim Re^{-1/4} \left(1 + \frac{\omega \ell}{u}\right)^{3/4} . \qquad (7.48)$$

The limit of Eq. (7.48) for $\omega \to 0$,

$$\frac{1}{2}f \sim Re^{-1/4} , \qquad (7.49)$$

is known to correlate turbulent data on flat plates.

Now, let heat transfer be expressed in terms of *local thermal* scale η_θ,

$$Nu \sim \frac{\ell}{\eta_\theta} , \qquad (7.50)$$

Nu being the usual Nusselt number. The ratio of Eqs. (7.44) and (7.50) gives,

$$\frac{Nu}{Re} \sim \frac{1}{2}f \left(\frac{\eta}{\eta_\theta}\right) \qquad (7.51)$$

which, in terms of Eq. (7.39), can be rearranged for the Stanton number,

$$StPr^{2/3} = \frac{1}{2}f \left[1 + \frac{\omega \lambda}{u} \left(Pr^{1/3} - 1\right)\right]^{1/3} , \qquad (7.52)$$

or, in terms of Eqs. (7.39) and (7.48),

$$Nu \sim Re^{3/4} \left(1 + \frac{\omega \ell}{u}\right)^{3/4} \left[1 + \frac{\omega \lambda}{u} \left(Pr^{1/3} - 1\right)\right]^{1/3} Pr^{1/3} . \qquad (7.53)$$

Finally, recalling Eq. (7.12) and assuming $U \gg U_0$, Eq. (7.53) is rearranged as

$$Nu \sim \overline{Re}^{3/4} \left(1 + \frac{U_0}{U}\right)^{3/4} \left(1 + \frac{\omega \ell}{U}\right)^{3/4} \times$$

$$\left[1 + \frac{\omega \ell}{U}(\overline{Re})^{-1/2} \left(Pr^{1/3} - 1\right)\right]^{1/3} Pr^{1/3} , \qquad (7.54)$$

or, for $Pr = 1$,

$$Nu \sim \overline{Re}^{3/4} \left(1 + \frac{U_0}{U}\right)^{3/4} \left(1 + \frac{\omega\ell}{U}\right)^{3/4} , \qquad (7.55)$$

where $\overline{Re} = U\ell/\nu$.

Using this model in the next section, we correlate heat transfer data on pulse combustor tailpipes.

7.5   PULSE COMBUSTION

Enhanced rates of heat transfer in pulse combustor tailpipes result from large flow oscillations (caused by the acoustic resonance of combustor) superimposed on turbulent velocity fluctuations. The literature on the effect of flow oscillations on heat transfer rates in turbulent flows is rather controversial. Heat-transfer rates in pulse combustor tailpipes have been found to vary from 70% less (Alhaddad and Coulman, 1982), to 240% greater (Hanby, 1969), than those of steady flow at the same mean Reynolds number. Other oscillating flows have shown decreases in the heat transfer coefficient of up to 20% (Liao et al., 1985) and increases of up to a factor of 5 (Galitseyskiy and Ryzhov, 1977) over steady flow conditions. Part of these inconsistencies may be explained by the greatly different flow conditions of the studies. Also, these studies lacked systematic variation of the important flow parameters, and many were conducted at frequencies much lower than the 45 to 200 Hz range, typical of pulse combustors. The effect of flow oscillations on pulse combustor tailpipe heat transfer has recently been clarified by the experimental study of Dec and Keller (1989). They systematically varied the important flow parameters such as frequency, velocity oscillation amplitude and mean flow rate and accessed the effects of these parameters on the heat transfer.

The heat transfer model introduced in the preceding section is used here for a correlation of the experimental data of Dec and Keller. Previous modeling of oscillating flow heat transfer has been based on quasi-steady assumptions, which are valid only in flows with oscillation frequencies lower than those typical of pulse combustors, and results in a heat-transfer correlation which is independent of frequency. The heat-transfer data of Dec and Keller , which was obtained over a range of typical pulse combustor frequencies (54 to 101 Hz), clearly shows a frequency dependence that cannot be explained by a quasi-steady model.

Now, with some constants, replacing Eq. (7.55) with an equality, and following Arpacı et al. (1993), we have

$$Nu = C_0 \overline{Re}^{3/4} \left(1 + C_1 \frac{U_0}{U} + C_3 \frac{\omega \ell}{U} + C_4 \frac{U_0 \omega \ell}{U^2} \right)^{3/4} . \tag{7.56}$$

Experimental results show that

$$C_3 \ll C_1, C_4 . \tag{7.57}$$

After dropping $C_3(\omega \ell / U)$, introducing $C_2 = C_4/C_1$, Eq. (7.56) becomes

$$Nu = C_0 \overline{Re}^{3/4} \left[1 + C_1 \frac{U_0}{U} \left(1 + C_2 \frac{\omega \ell}{U} \right) \right]^{3/4} . \tag{7.58}$$

Furthermore, let the integral scale ℓ be the hydraulic diameter D of the tailpipe[3] and, following Dec and Keller, assume $\omega \ell = (\omega - \omega_0)D$, ω_0 denoting the pulsating frequency below which a quasi-steady model independent of frequency yields correct results. Numerical constants determined by a least-squares fit of experimental data are $C_0 = 0.0291$, $C_1 = 0.1762$, $C_2 = 8.483$ and $f_0 = \omega_0/2\pi = 45.44$ with a variance of 3.23. With these values, Eq. (7.58) becomes

$$Nu = 0.0291 \overline{Re}^{3/4} \left\{ 1 + 0.176 \frac{U_0}{U} \left[1 + 8.48 \frac{(f - 45)D}{U} \right] \right\}^{3/4} . \tag{7.59}$$

Fig. 7.2 shows the results of Eq. (7.59). In view of the complexity of the problem, the agreement between the data and the correlation is remarkably good. For $f \leq f_0$ Eq. (7.59) reduces to a quasi-steady model,

$$Nu = 0.029 \overline{Re}^{3/4} \left(1 + 0.21 \frac{U_0}{U} \right)^{3/4} . \tag{7.60}$$

For steady flow, $U_0 = 0$ and Eq. (7.60) reduces to a steady model,

$$Nu = 0.029 \overline{Re}^{3/4} \tag{7.61}$$

which is quite close to the Colburn correlation for heat transfer in steady turbulent flows with $Pr = 1$,

---

[3]The natural frequency of a resonator depends on the length of the orifice or tailpipe (see the next section for a brief review). Using a sectional tailpipe, Dec and Keller are able to operate a pulse combustor over a frequency range. In the correlation given by Eq. (7.59), frequency is a measured input and the length scale involved with $\omega \ell / U$ is immaterial.

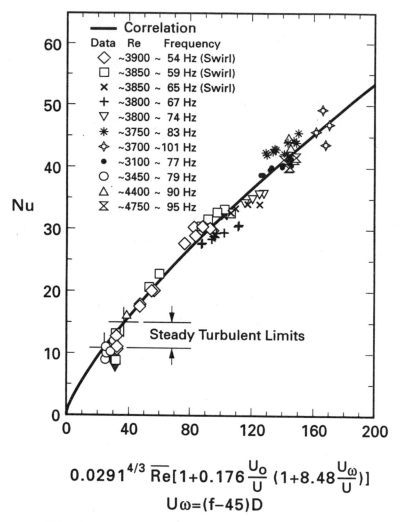

$$0.0291^{4/3}\,\overline{Re}\,[1+0.176\frac{U_o}{U}\,(1+8.48\frac{U\omega}{U})]$$

$$U\omega = (f-45)D$$

FIGURE 7.2: A correlation of the experimental data with Eq. (7.59).

$$Nu = 0.023\overline{Re}^{0.8}\ . \tag{7.62}$$

Thus, the correlation given by Eq. (7.59) not only collapses and correlates all the experimental data, but it also reduces to the correct limits for $f \to f_0$ and $U_0 \to 0$. However, before any claim can be made on any universal nature of this correlation, or the values of the constants

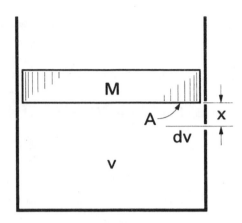

FIGURE 7.3: Piston-Cylinder assembly.

presented here, a more extensive data base than the one presently available is needed.

In concluding, we state that the intuitive Kolmogorov ideas on the microsturcture of turbulence are universal and apply to all turbulent flows, but the well-known forms of these scales are for steady (on the mean) isothermal flows. In the present chapter, we introduced microscales appropriate for oscillating (on the mean) turbulent flows, and modeled skin friction and heat transfer in terms of these scales. The special case of the heat transfer model for $Pr = 1$ is shown to correlate the recent experimental literature on the oscillating turbulent flow in the tailpipe of a pulse combustor. The maximum enhancement turns out to be about 300% within the range of the physical parameters considered.

The next section deals with a brief review of the Helmholtz resonator.

7.6   HELMHOLTZ RESONATOR

Consider a piston-cylinder assembly (Fig. 7.3). Neglecting friction, assume that the piston steadily oscillates in the cylinder. For a mass of the piston considerably greater than the air confined to the cylinder, neglecting the inertia of air, assume the air to act as an elastic medium. Then, a balance between the kinetic energy of the piston and the potential energy of the air yields the natural frequency of the piston oscillation.

The kinetic energy increase of the piston displaced by a distance x is

$$\Delta U_K = \frac{1}{2} M \left(\frac{dx}{dt} \right)^2 ,$$
(7.63)

where M is the mass of the piston. The potential energy decrease of, or the work done on, the air of volume $d\mathcal{V}$ is

$$\Delta U_P = \frac{1}{2} \sigma \epsilon \, d\mathcal{V} ,$$
(7.64)

where σ is a normal stress and ϵ is the strain rate. In terms of $\sigma = p$ and $\epsilon = d\mathcal{V}/\mathcal{V}$,

$$\Delta U_P = \frac{1}{2} p \left(\frac{d\mathcal{V}}{\mathcal{V}} \right) d\mathcal{V}$$
(7.65)

and, in view of $p = \rho a^2$ and $d\mathcal{V} = Ax$,

$$\Delta U_P = \frac{1}{2} \rho a^2 \frac{A^2 x^2}{\mathcal{V}} ,$$
(7.66)

a being the speed of sound and A the cross-sectional area of the piston. Then, from energy balance,

$$\Delta(U_K + U_P) = 0 ,$$
(7.67)

we have, following a differentiation converting Eq. (7.67) to a force balance,

$$\frac{d^2 x}{dt^2} + \left(\frac{\rho a^2 A^2}{M \mathcal{V}} \right) x = 0$$
(7.68)

which describes a harmonic motion with natural frequency

$$\omega^2 = \frac{\rho a^2 A^2}{M \mathcal{V}} .$$
(7.69)

Now consider the Helmholtz resonator consisting of a chamber-tail-pipe assembly (Fig. 7.4). Eq. (7.69) provides a first order approximation for the frequency of this assembly. Now, viewing the foregoing energy balance as a balance between the potential energy of air in the chamber

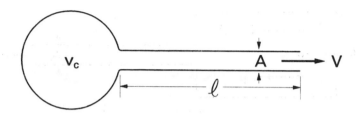

FIGURE 7.4: Helmholtz resonator.

and the kinetic energy of air in the tailpipe, assuming $V = V_c$, $M = \rho V_P$ and noting $V_P = A\ell$, we get from Eq. (7.69),

$$\omega_H = a\left(\frac{A}{\ell V_c}\right)^{1/2} \tag{7.70}$$

which shows the dependence of frequency on the length and cross-section of the tailpipe and the volume but not the shape of the chamber. Also, note in terms of V_P that

$$\frac{\omega_H \ell}{a} = \left(\frac{V_P}{V_C}\right)^{1/2} . \tag{7.71}$$

For a more accurate model including the effect of air inertia, consider the dynamics of one-dimensional waves in a resonator consisting of a cylindric chamber and a tailpipe (Fig. 7.5). Let the solution of the classical wave equation for pressure,

$$\frac{\partial^2 p}{\partial t^2} = a^2 \frac{\partial^2 p}{\partial x^2} , \tag{7.72}$$

be written, in terms of incident and reflected waves, as

$$p = p_I + p_R , \tag{7.73}$$

or,

$$p = C_I e^{i(\omega t + kx)} + C_R e^{i(\omega t - kx)} , \tag{7.74}$$

or,

$$p = C_I e^{i\omega t}\left(e^{ikx} + Re^{-ikx}\right) , \tag{7.75}$$

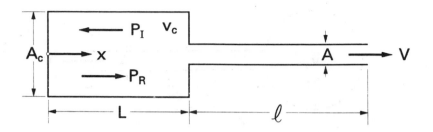

FIGURE 7.5: A one-dimensional resonator.

where $R = C_R/C_I$ and $\omega = ka$. Also, in view of the linearized momentum,

$$\frac{\partial u}{\partial t} + \frac{1}{\rho}\frac{\partial p}{\partial x} = 0 , \qquad (7.76)$$

we have

$$u = -\frac{C_I}{\rho a}e^{i\omega t}\left(e^{ikx} - Re^{-ikx}\right) , \qquad (7.77)$$

and the acoustic impedance,

$$Z(x) = \frac{p}{|u|A_c} = \frac{\text{Pressure}}{\text{Flow rate}} , \qquad (7.78)$$

which gives, in terms of Eqs. (7.75) and (7.77),

$$Z(x) = \frac{\rho a}{A_c}\left(\frac{e^{ikx} + Re^{-ikx}}{e^{ikx} - Re^{-ikx}}\right) . \qquad (7.79)$$

For $x = 0$, $Z(0) = \infty$ because of $u(0) = 0$. This condition yields $R = 1$. Then,

$$Z(x) = i\left(\frac{\rho a}{A_c}\right)\cot kx \qquad (7.80)$$

and

$$Z(L) = i\left(\frac{\rho a}{A_c}\right)\cot kL , \qquad (7.81)$$

where L is the length of the chamber. In addition, from the mass oscillating in the tailpipe,

$$Z(L) = \frac{F}{A^2|u|} = \frac{m\left|\dfrac{du}{dt}\right|}{A^2|u|} = i\omega\frac{m}{A^2} , \tag{7.82}$$

or, noting $m = \rho A\ell$,

$$Z(L) = i\omega\left(\frac{\rho\ell}{A}\right) . \tag{7.83}$$

Then, from the equality between Eqs. (7.81) and (7.83),

$$\frac{\omega\ell}{a} = \left(\frac{A}{A_c}\right)\cot kL . \tag{7.84}$$

For $kL < 1$,

$$\cot kL = \frac{1}{kL} - \frac{1}{3}kL + \mathcal{O}(kL)^2 \tag{7.85}$$

and, Eq. (7.84) becomes

$$\frac{\omega\ell}{a} = \left(\frac{\mathcal{V}_P}{\mathcal{V}_C}\right)^{1/2}\left[1 - \frac{1}{3}\left(\frac{\omega\ell}{a}\right)^2\left(\frac{L}{\ell}\right)^2\right]^{1/2} , \tag{7.86}$$

or, following an arrangement,

$$\frac{\omega\ell}{a} = \frac{(\mathcal{V}_P/\mathcal{V}_C)^{1/2}}{\left[1 + \frac{1}{3}(\mathcal{V}_P/\mathcal{V}_C)(L/\ell)^2\right]^{1/2}} , \tag{7.87}$$

or, in terms of the Helmholtz frequency (recall Eq. 7.71),

$$\frac{\omega\ell}{a} = \frac{\omega_H\ell/a}{\left[1 + \frac{1}{3}(\omega_H L/a)^2\right]^{1/2}} . \tag{7.88}$$

In general, $\mathcal{V}_P \ll \mathcal{V}_C$, $\ell \sim L$, and

$$\frac{\omega_H L}{a} = \left(\frac{\mathcal{V}_P}{\mathcal{V}_C}\right)^{1/2}\frac{L}{\ell} \ll 1 \tag{7.89}$$

which demonstrates the fact that the correction on the Helmholtz frequency resulting from air inertia is quite small.

PROBLEMS

7-1) Develop the phase averaged equations for thermal dynamics leading to Eq. (7.26).

7-2) Develop the microscales of buoyancy-driven oscillating flows.

CHAPTER 8

TWO-PHASE FILMS

The literature on heat transfer in two-phase films goes back to Nusselt (1916) who obtained in terms of a Rayleigh number based on latent heat

$$Nu = f(Ra_2) \,, \tag{8.1}$$

or, explicitly,

$$Nu \sim Ra_2^{1/4} \tag{8.2}$$

for laminar films. This relation ignores the effect of inertial force which is characterized by the two-phase Prandtl number,

$$Nu = f(Ra_2, Pr_2) \,. \tag{8.3}$$

This relation is implicit in the similarity solution of Sparrow and Gregg (1959) for the laminar film condensation of saturated vapor next to a vertical isothermal wall. Koh, et al. (1961) included later the effect of vapor flow on the same problem.

The foregoing dimensionless numbers result from the coupling of appropriate (inertial, viscous and buoyant) forces of the momentum balance with enthalpy flow and conduction terms of the thermal energy balance. However, when more than two forces are involved in single-phase or two-phase heat transfer, a dimensionless number (say Π_2 for two-phase) can be found which includes a combination of Ra_2 and Pr_2, as well as Ra and Pr, such as

$$\Pi_2 \sim \frac{Ra_2 + Ra}{1 + (Pr_2 + Pr)^{-1}} \,, \tag{8.4}$$

161

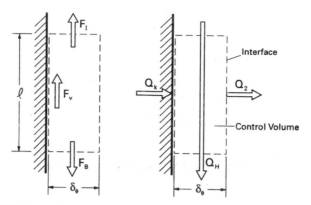

FIGURE 8.1: Control volume for (a) momentum and (b) energy.

and leads to

$$Nu = f(\Pi_2) \ . \tag{8.5}$$

The objective of this chapter is correlation of two-phase turbulent films in terms of Eq. (8.5). The fundamental significance of Π_2 for two-phase films appears to be sofar overlooked except for the recent work of Arpacı and Larsen (1988), Arpacı, et al. (1990), Arpacı and Kao (1996).

8.1   A DIMENSIONLESS NUMBER (Π_2)

Consider the control volume shown in Fig. 8.1. The balance among body and surface forces acting on this control volume yields

$$F_I + F_V \sim F_B \ , \tag{8.6}$$

F_I being the inertial force, F_V the viscous force and F_B the buoyant force. Interphase force F_2 is neglected in Eq. (8.6) since, in most practical situations,

$$\frac{F_2}{F_I} = \frac{\dot{m}_2^2 |(1/\rho_1) - (1/\rho_2)|}{\dot{m}_I^2 (1/\rho_1)} \ll 1 \tag{8.7}$$

where ρ_1 and ρ_2 indicating the densities of Phase 1 and Phase 2, respectively.

The balance of thermal energy for the same control volume gives

$$Q_2 + Q_H \sim Q_K \ , \tag{8.8}$$

Q_2 being the enthalpy flow across the two-phase interface, Q_H the longitudinal net enthalpy flow, and Q_K the conduction. The reasons for retaining Q_2, while neglecting F_2, will be clear later.

A dimensionless number resulting from Eq. (8.6) is

$$\frac{F_B}{F_I + F_V}, \tag{8.9}$$

or,

$$\frac{F_B/F_V}{F_I/F_V + 1} = \left(\frac{F_V/F_I}{1 + F_V/F_I}\right)\frac{F_B}{F_V}. \tag{8.10}$$

Also, a dimensionless number associated with Eq. (8.8) is

$$\frac{Q_2}{Q_K} + \frac{Q_H}{Q_K} = \frac{Q_2}{Q_K}\left(1 + \frac{Q_H}{Q_2}\right). \tag{8.11}$$

On dimensional grounds,

$$\frac{F_B}{F_V} \sim \frac{g(\Delta\rho)\ell^2}{\mu V} \; , \quad \frac{F_I}{F_V} \sim \frac{\rho V \ell}{\mu} \; , \quad \frac{Q_2}{Q_K} \sim \frac{\rho V \ell h_{fg}}{k\Delta T} \; , \quad \frac{Q_H}{Q_2} \sim \frac{c_p \Delta T}{h_{fg}} \; ,$$

where ℓ is a characteristic length and the rest of the notation is conventional. However, the foregoing nondimensionalizations in terms of a characteristic velocity V are not appropriate for buoyancy-driven flows. Velocity is now a dependent variable and should not appear in the dimensionless numbers describing these flows. Accordingly, one needs to combine Eq. (8.11) with Eq. (8.10) for a result independent of V. The only combination which eliminates velocity between Eq. (8.10) and (8.11) is

$$\Pi_2 \sim \frac{(F_B/F_V)(Q_2/Q_K + Q_H/Q_K)}{(F_I/F_V)(Q_2/Q_K + Q_H/Q_K)^{-1} + 1}, \tag{8.12}$$

or

$$\Pi_2 \sim \frac{Ra_2 + Ra}{1 + (Pr_2 + Pr)^{-1}} \tag{8.13}$$

which may be rearranged in terms of the Jacob number,

$$Ja = \frac{c_p \Delta T}{h_{fg}} \sim \frac{Q_H}{Q_2}, \tag{8.14}$$

as

$$\Pi_2 \sim \frac{Ra_2(1 + Ja)}{1 + [Pr_2(1 + Ja)]^{-1}} = \left[\frac{Pr_2(1 + Ja)}{1 + Pr_2(1 + Ja)}\right] Ra_2(1 + Ja) , \quad (8.15)$$

where

$$Ra_2 \sim \left(\frac{F_B}{F_V}\right)\left(\frac{Q_2}{Q_K}\right) \sim \frac{g(\Delta\rho)\rho h_{fg}\ell^3}{\mu k(\Delta T)} \quad (8.16)$$

and

$$Pr_2 \sim \left(\frac{F_V}{F_I}\right)\left(\frac{Q_2}{Q_K}\right) \sim \frac{\mu h_{fg}}{k\Delta T} . \quad (8.17)$$

For notational convenience, letting

$$R_2 \sim Ra_2(1 + Ja) , \quad P_2 \sim Pr_2(1 + Ja) , \quad (8.18)$$

Eq. (8.15) may in form be reduced to

$$\Pi_2 \sim \frac{R_2}{1 + P_2^{-1}} = \left(\frac{P_2}{1 + P_2}\right) R_2 . \quad (8.19)$$

The heat transfer across a two-phase film is then represented by

$$Nu = f(\Pi_2) . \quad (8.20)$$

Clearly, in most cases,

$$Ja \ll 1 \quad (8.21)$$

and Eq. (8.19) is reduced to

$$\Pi_2 \sim \frac{Ra_2}{1 + Pr_2^{-1}} = \left(\frac{Pr_2}{1 + Pr_2}\right) Ra_2 \quad (8.22)$$

whose two limits, respectively obtained from the first and second forms of Π_2, are

$$\lim_{Pr_2 \to \infty} \Pi_2 = Ra_2 \quad (8.23)$$

and

$$\lim_{Pr_2 \to 0} \Pi_2 = Ra_2 Pr_2 . \quad (8.24)$$

Note that a "two-phase specific heat,"

$$c_{p2} = h_{fg}/\Delta T \ , \tag{8.25}$$

may be defined as the natural limit of $(\partial h/\partial T)_p$, ΔT being the temperature jump across interface. In terms of this definition,

$$\alpha_2 = \frac{k}{\rho c_{p2}} \ , \quad Ra_2 = \frac{g}{\nu \alpha_2} \left(\frac{\Delta \rho}{\rho}\right) \ell^3 \ , \quad Pr_2 = \frac{\nu}{\alpha_2} \ , \tag{8.26}$$

and Ra_2 and Pr_2 assume their conventional forms. Here, for customary reasons only, Eqs. (8.16) and (8.17) are rearranged by Q_H,

$$Ra_2 \sim \left(\frac{F_B}{F_V}\right) \left(\frac{Q_H}{Q_K}\right) \left(\frac{Q_2}{Q_H}\right) \tag{8.27}$$

and

$$Pr_2 \sim \left(\frac{F_V}{F_I}\right) \left(\frac{Q_H}{Q_K}\right) \left(\frac{Q_2}{Q_H}\right) \tag{8.28}$$

which, in terms of the usual Rayleigh and Prandtl numbers,

$$Ra \sim \left(\frac{F_B}{F_V}\right) \left(\frac{Q_H}{Q_K}\right) \sim \frac{g}{\nu \alpha} \left(\frac{\Delta \rho}{\rho}\right) \ell^3 \ , \tag{8.29}$$

and

$$Pr \sim \left(\frac{F_V}{F_I}\right) \left(\frac{Q_H}{Q_K}\right) \sim \frac{\nu}{\alpha} \tag{8.30}$$

become

$$Ra_2 = Ra/Ja \ , \quad Pr_2 = Pr/Ja \ . \tag{8.31}$$

The next two sections develop two explicit forms of Eq. (8.20) corresponding to laminar and turbulent films of two-phase flows, respectively.

8.2   LAMINAR TWO-PHASE

In terms of a flow scale ℓ (or x) and a diffusion scale δ, an explicit dimensionless form of Eq. (8.6) is

$$u\frac{u}{\ell} + \nu\frac{u}{\delta^2} \sim g\left(\frac{\Delta \rho}{\rho}\right) \ , \tag{8.32}$$

and an explicit dimensionless form of Eq. (8.8) for $Ja \ll 1$ is

$$\rho \left(\frac{u\delta_\theta}{\ell} \right) h_{fg} \sim k \frac{\Delta T}{\delta_\theta} \, , \qquad (8.33)$$

where $u\delta_\theta/\ell$ is the interface velocity of transversal mass flow expressed in terms of the longitudinal mass flow.

Noting that the thickness of momentum and thermal boundary layers is about the same,

$$\delta \sim \delta_\theta \, , \qquad (8.34)$$

and rearranging Eq. (8.32) in terms of Eq. (8.34) yields

$$\frac{u}{\delta_\theta^2} \left(1 + \frac{u\delta_\theta^2}{\nu\ell} \right) \sim \frac{g}{\nu} \left(\frac{\Delta\rho}{\rho} \right) \, . \qquad (8.35)$$

Separately, Eq. (8.33) gives

$$\frac{u\delta_\theta^2}{\ell} \sim \frac{k\Delta T}{\rho h_{fg}} \qquad (8.36)$$

which may be rearranged in terms of Eqs. (8.25) and (8.26) as

$$\frac{u\delta_\theta^2}{\ell} \sim \frac{k}{\rho c_{p2}} = \alpha_2 \, , \qquad (8.37)$$

or,

$$u \sim \frac{\alpha_2 \ell}{\delta_\theta^2} \, . \qquad (8.38)$$

Insertion of Eq. (8.38) into Eq. (8.35) leads to

$$\frac{\ell}{\delta_\theta^4} \sim \frac{(g/\nu\alpha_2)(\Delta\rho/\rho)}{1 + Pr_2^{-1}} \, , \qquad (8.39)$$

or, in terms of Eq. (8.22), to

$$\frac{\ell}{\delta_\theta} \sim \Pi_2^{1/4} \sim Nu \, . \qquad (8.40)$$

Two limits of Eq. (8.40) are

$$\lim_{Pr_2 \to \infty} Nu \sim Ra_2^{1/4} \qquad (8.41)$$

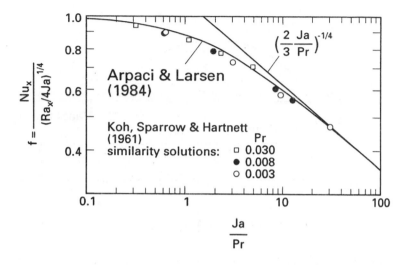

FIGURE 8.2: Departure from Nusselt ($f = 1$) for liquid metals.

and

$$\lim_{Pr_2 \to 0} Nu \sim (Ra_2 Pr_2)^{1/4} . \tag{8.42}$$

Clearly, the development including the effect of sensible heat leads to Π_2 in terms of Eq. (8.19) rather than Eq. (8.22). Next, we consider a specific example explicitly demonstrating the existence of Π_2.

A straightforward integral approach by Arpacı and Larsen (1984), applied here to laminar film condensation of saturated vapor next to an isothermal vertical wall, yields for the local heat transfer,

$$Nu_x = \left[\frac{1}{4} (\Pi_2)_x \right]^{1/4} , \quad (\Pi_2)_x = \frac{(Ra_2)_x}{1 + \frac{2}{3} Pr_2^{-1}} . \tag{8.43}$$

These results agree well with the similarity solution of Koh, et al. (1961), show also the deviation, $f = Nu_x / [(Ra_2)_x / 4]^{1/4}$, from the Nusselt relation due to inertial effects (Fig. 8.2).

In addition, consider an integral approach to dropwise film boiling on an isothermal horizontal plate. For a first order model, let the geometry of actual drops (which is determined by gravity, surface tension and pressure distribution) be approximated by a hemisphere (Fig. 8.3).

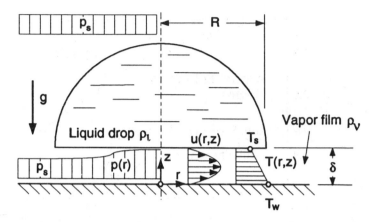

FIGURE 8.3: Idealized drop evaporation.

The liquid is at rest and isothermal at saturation temperature, T_s. The heating surface is isothermal at T_w which exceeds the Leidenfrost temperature.

A quasi-steady integral formulation for the vapor film is

$$\frac{1}{r}\frac{\partial}{\partial r}\left(r\int_0^\delta \rho u\, dz\right) - w_\delta = 0 \ , \tag{8.44}$$

$$\frac{1}{r}\frac{\partial}{\partial r}\left(r\int_0^\delta \rho u^2\, dz\right) - u(r,\delta)w_\delta = -\delta\frac{\partial p}{\partial r} + \tau_{zr}|_{z=\delta} - \tau_{zr}|_{z=0} \ , \tag{8.45}$$

$$\frac{1}{r}\frac{\partial}{\partial r}\left(r\int_0^\delta \rho c_p u T\, dz\right) - c_p T(r,\delta)w_\delta = -q_z|_{z=\delta} + q_z|_{z=0} \ , \tag{8.46}$$

where w_δ denotes the evaporating mass flow per unit area, and

$$u(r,0) = 0 \ ; \quad u(r,\delta) = 0 \ ; \quad T(r,0) = T_w \ ; \quad T(r,\delta) = T_s \tag{8.47}$$

$$\tau_{zr} = \mu\frac{\partial u}{\partial z} \ ; \quad q_z = -k\frac{\partial T}{\partial z} \ . \tag{8.48}$$

The balance of thermal energy and the liquid-vapor interface is

$$h_{fg}w_\delta = -k\frac{\partial T(r,z)}{\partial z} \ , \tag{8.49}$$

where h_{fg} denotes the heat of evaporation.

For a quasi-developed solution, ignoring the left side of Eq. (8.46), assume a linear temperature distribution,

$$\frac{T - T_s}{T_w - T_s} = 1 - \frac{z}{\delta} \, , \tag{8.50}$$

also, neglecting the radial variation of the vapor film, assume a parabolic velocity distribution,

$$\frac{u}{u_m} = 6\frac{z}{\delta}\left(1 - \frac{z}{\delta}\right) \, , \tag{8.51}$$

where $u_m(r)$ denotes the mean vapor velocity. A quasi-developed formulation of both the momentum and energy balances is justified only by the Reynolds analogy provided that the Prandtl number is of order unity and the pressure gradient is zero.

Using Eqs. (8.50) and (8.51), eliminating w_δ by Eq. (8.49), Eqs. (8.44) and (8.45) are reduced to

$$\frac{1}{r}\frac{\partial}{\partial r}(ru_m) = \frac{\alpha Ja}{\delta} \, , \tag{8.52}$$

$$\frac{6}{5}\frac{\delta}{r}\frac{\partial}{\partial r}(ru_m^2) = -\frac{\delta}{\rho}\frac{\partial p}{\partial r} - \frac{12\nu u_m}{\delta} \, , \tag{8.53}$$

where $Ja = c_p(T_w - T_s)/h_{fg}$ denotes the Jacob number. Using symmetry and denoting the pressure outside the drop by p_s, boundary conditions may be stated as

$$u_m(0) = 0 \, ; \quad \frac{dp(0)}{dr} = 0 \, ; \quad p(R) = p_s \, ; \quad \delta = \text{Const.} \tag{8.54}$$

Introducing $u_m = \alpha r Ja/2\delta^2$, obtained from Eq. (8.52), into Eq. (8.53) and integrating once yields

$$p(r) - p_s = 3\mu\alpha Ja\left(\frac{R^2}{\delta^4}\right)\left(1 + \frac{3}{20}\frac{Ja}{Pr}\right)\left[1 - \left(\frac{r}{R}\right)^2\right] \, . \tag{8.55}$$

The limit of Eq. (8.55) for $Ja/Pr \to 0$ is available in the literature (Wachters, Bonne, and van Nouhuis, 1966; Avedisian, Ioffredo, and O'Connor, 1984).

Now, the vertical balance of momentum for the drop, ignoring unsteadiness and momentum flow due to evaporation, gives

$$0 = \int_0^R 2\pi r \left[p(r) - p_s \right] \, dr - (\rho_\ell - \rho) g \frac{2}{3} \pi R^3 \,, \tag{8.56}$$

where the subscript ℓ refers to liquid. Inserting Eq. (8.55) into Eq. (8.56), and using $D = 2R$ as the characteristic length, yields the film thickness

$$\delta = \left[\frac{9}{8} \frac{\alpha \mu J a D}{g(\rho_\ell - \rho)} \left(1 + \frac{3}{20} \frac{Ja}{Pr} \right) \right]^{1/4} \,. \tag{8.57}$$

It also can be shown that the droplet evaporation lifetime is

$$\tau_e = \frac{\delta_o D_o \rho_\ell h_{fg}}{5 k \Delta T} \,, \tag{8.58}$$

where subscript o indicating droplet initial state. Finally, introducing the mean instantaneous heat transfer coefficient,

$$\overline{h}(t) = \frac{1}{T_w - T_s} \frac{1}{\pi R^2} \int_0^R 2\pi r q_y(r, 0) \, dr \,, \tag{8.59}$$

and using Eq. (8.57), yields

$$Nu = \frac{\overline{h}(t) D}{k} = \left[\frac{8}{9} \frac{g(\rho_\ell - \rho) D^3}{\mu \alpha Ja \left(1 + \frac{3}{20} \frac{Ja}{Pr} \right)} \right]^{1/4} \tag{8.60}$$

which may be rearranged in terms of

$$\Pi_2 = \frac{Ra/Ja}{1 + \frac{3}{20} Ja/Pr} \tag{8.61}$$

as

$$Nu = \left(\frac{8}{9} \Pi_2 \right)^{1/4} \,. \tag{8.62}$$

There are experimental literature which can also be rearranged in terms of Π_2. Following Bromley (1950), existing correlations for film boiling on an immersed plate, cylinder or sphere assume the form of the correlation for film condensation. The difference between the two cases is that, in condensation, liquid film flows downward and slightly influenced by the adjacent vapor while, in boiling, vapor film flows upward and strongly influenced by the adjacent liquid. However, this difference only makes the vapor film thicker which lead to constants smaller than those involved with condensation. The literature gives, in terms of the limit of Π_2 for $Pr_2 \to \infty$ obtained from Eq. (8.22),

$$\bar{h} = C \left[\frac{(\rho_\ell - \rho_v)gh'_{fg}k_v^3}{\nu_v(T_w - T_{\text{sat}})L}\right]^{1/4} , \qquad (8.63)$$

L being the height of a vertical plate, $L = D$ for sphere and cylinder, and $h'_{fg} = h_{fg} + C_0 c_{pv}(T_w - T_{\text{sat}})$. Here C is 0.71 for vertical plate, 0.62 for horizontal cylinder and 0.67 for sphere (Frederking and Daniels, 1966), respectively, and $C_0 = 0.35$ is assumed for all geometries. Clearly, Eq. (8.63) neglects the inertial effect and, in view of Eq. (8.18), can be represented by

$$Nu = C_1 R_2^{1/4} , \qquad R_2 = Ra_2(1 + C_0 Ja) . \qquad (8.64)$$

The next section is devoted to turbulent films associated with two-phase flows. Since the integral methods are no longer suitable, the development is based on dimensional arguments leading to appropriate microscales and, depending on these scales, heat transfer in terms of Π_2. Experimental support is provided by the existing literature.

8.3   TURBULENT TWO-PHASE

Let the instantaneous velocity and temperature of a buoyancy driven flow be decomposed into temporal means and fluctuations,

$$U_i + u_i , \quad \Theta + \theta ,$$

and let U_i and Θ be statistically steady. Then, the balance of the mean kinetic energy of homogeneous velocity fluctuations gives

$$\mathcal{B} = \mathcal{P} + \epsilon \qquad (8.65)$$

where

$$\mathcal{B} = -g_i \overline{u_i \theta}/\Theta_0 \qquad (8.66)$$

is the buoyant production, g_i being the vector acceleration of gravity and Θ_0 a characteristic temperature for isobaric ambient,

$$\mathcal{P} = -\overline{u_i u_j} S_{ij} \tag{8.67}$$

is the inertial production, S_{ij} being the rate of the mean strain, and

$$\epsilon = -2\nu \overline{s_{ij} s_{ij}} \tag{8.68}$$

is the dissipation of turbulent energy, s_{ij} being the rate of fluctuating strain.

Also, for $Ja \ll 1$, the balance of the root mean square of homogeneous thermal fluctuations involving phase change yields

$$(\mathcal{P}_\theta)_2 = (\epsilon_\theta)_2 \tag{8.69}$$

where

$$(\mathcal{P}_\theta)_2 = -\overline{u_i \theta} \frac{\partial \Theta}{\partial x_i} \tag{8.70}$$

is the mean thermal production, and

$$(\epsilon_\theta)_2 = \alpha_2 \overline{\left(\frac{\partial \theta}{\partial x_i}\right)\left(\frac{\partial \theta}{\partial x_i}\right)} \tag{8.71}$$

is the thermal dissipation, $\alpha_2 = k/\rho c_{p2}$ and $c_{p2} = h_{fg}/\Delta T$ being already defined by Eqs. (8.26) and (8.25).

On dimensional grounds, Eq. (8.65) yields

$$\mathcal{B} \sim \frac{u^3}{\ell} + \nu \frac{u^2}{\lambda^2} \tag{8.72}$$

where u is the rms value of velocity fluctuations, ℓ is an integral scale and λ is the Taylor scale, and Eq. (8.69) leads to

$$(\mathcal{P}_\theta)_2 \sim u \frac{\theta^2}{\ell} \sim \alpha_2 \frac{\theta^2}{\lambda_\theta^2} \sim (\epsilon_\theta)_2 , \tag{8.73}$$

λ_θ being a thermal Taylor scale characterizing the film thickness.

Assuming in a manner similar to Eq. (8.34),

$$\lambda \sim \lambda_\theta , \tag{8.74}$$

rearranging Eq. (8.72) in terms of Eq. (8.73), and solving the result for λ_θ, yields

$$\lambda_\theta \sim \ell^{1/3}(1 + Pr_2)^{1/6}\left(\frac{\alpha_2^3}{B}\right)^{1/6} = \ell^{1/3}\left(1 + \frac{1}{Pr_2}\right)^{1/6}\left(\frac{\nu\alpha_2^2}{B}\right)^{1/6}.\quad(8.75)$$

Although λ_θ and ℓ are two different scales, let both approach the scale η_θ of the isotropic flow as the limit of the homogeneous flow,

$$\left(\frac{\lambda_\theta}{\ell}\right) \to \eta_\theta .\quad(8.76)$$

Then, Eq. (8.75) is reduced to a thermal Kolmogorov scale for two-phase turbulent flows,

$$\eta_\theta \sim (1 + Pr_2)^{1/4}\left(\frac{\alpha_2^3}{B}\right)^{1/4} = \left(1 + \frac{1}{Pr_2}\right)^{1/4}\left(\frac{\nu\alpha_2^2}{B}\right)^{1/4}\quad(8.77)$$

and its two limits

$$\lim_{Pr_2 \to 0}\eta_\theta \sim \left(\frac{\alpha_2^3}{B}\right)^{1/4} ,\qquad \lim_{Pr_2 \to \infty}\eta_\theta \sim \left(\frac{\nu\alpha_2^2}{B}\right)^{1/4}\quad(8.78)$$

which formally are Oboukhov-Corrsin and Batchelor scales, respectively. For $\alpha_2 = \alpha$ and $Pr_2 = Pr$, Eq. (8.77) is reduced to a scale discovered by Arpacı (1986) for single-phase flows.

As demonstrated in the development leading to Π_2, a dimensionless number appropriate for buoyancy driven flows cannot depend on velocity which is a dependent variable. Since B depends on velocity, Eq. (8.77) cannot be the ultimate form of the microscales for buoyancy driven flows. To eliminate velocity, consider B in terms of the buoyant force $\mathcal{F}$. On dimensional grounds,

$$B \sim u\mathcal{F}\quad(8.79)$$

which may be rearranged in terms of the isotropic velocity obtained from Eq. (8.73),

$$u \sim \alpha_2/\eta_\theta ,\quad(8.80)$$

as

$$B \sim \alpha_2\mathcal{F}/\eta_\theta .\quad(8.81)$$

In terms of $\mathcal{F}$, Eq. (8.77) becomes

$$\eta_\theta \sim (1 + Pr_2)^{1/3} \left(\frac{\alpha_2^2}{\mathcal{F}}\right)^{1/3} = \left(1 + \frac{1}{Pr_2}\right)^{1/3} \left(\frac{\nu\alpha_2}{\mathcal{F}}\right)^{1/3} , \qquad (8.82)$$

which, introducing a "two-phase Rayleigh number,"

$$Ra_2 = \mathcal{F}\ell^3/\nu\alpha_2 , \qquad (8.83)$$

yields, relative to an integral scale,

$$\frac{\eta_\theta}{\ell} \sim (1 + Pr_2)^{1/3} (Ra_2 Pr_2)^{-1/3} = \left(1 + \frac{1}{Pr_2}\right)^{1/3} Ra_2^{-1/3} , \qquad (8.84)$$

or, in terms of Π_2,

$$(\ell/\eta_\theta) \sim \Pi_2^{1/3} \sim Nu , \qquad (8.85)$$

and

$$\lim_{Pr_2 \to 0} Nu \sim (Ra_2 Pr_2)^{1/3} , \qquad \lim_{Pr_2 \to \infty} Nu \sim (Ra_2)^{1/3} . \qquad (8.86)$$

Clearly, the development including the effect of sensible heat leads to Π_2 in terms of Eq. (8.19) rather than Eq. (8.22). The heat transfer relation suggested by Eq. (8.85) finds strong support in the single-phase literature on buoyancy driven turbulent flows (Arpacı, 1986, 1990a, 1995a, b).

For film boiling of cryogenic liquids at atmospheric pressure, correlating experimental data for liquid nitrogen, Frederking and Clark (1962) recommend,

$$\overline{h} = 0.15 \left[\frac{(\rho_\ell - \rho_v)gh'_{fg}k_v^2}{\nu_v(T_w - T_{sat})}\right]^{1/3} , \qquad \frac{(\rho_\ell - \rho_v)gh'_{fg}L^3}{k_v\nu_v(T_w - T_{sat})} > 5 \times 10^7 \quad (8.87)$$

where $h'_{fg} = h_{fg} + 0.5c_{pv}(T_w - T_{sat})$. This correlation can readily be expressed in terms of Eq. (8.18) leading to

$$Nu = 0.15R_2^{1/3} , \qquad R_2 = Ra_2(1 + 0.5Ja) > 5 \times 10^7 . \qquad (8.88)$$

Another correlation tested against data of film boiling heat transfer over an isothermal vertical surface in nitrogen (Kuriyama, 1987; Nishio et al., 1991; Suryanarayana and Merte, 1972), R-113 (Liaw and Dhir,

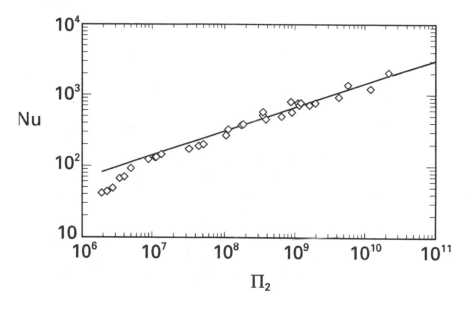

FIGURE 8.4: Nusselt number as a function of Π_2 for mercury.

1986), helium (Nishio and Chandratilleke, 1989), and water (Bui and Dhir, 1985) is obtained by Chu (1993) as

$$Nu = 0.148 Ra_2^{1/3} , \quad Ra_2 > 3.6 \times 10^6 . \tag{8.89}$$

Finally, the old mercury data of Lyon, et al. (1955) arranged in term of Π_2 leads to

$$Nu = 0.18 \Pi_2^{1/3} , \quad \Pi_2 = \frac{Ra_2 Pr_2}{1 + 0.03 Pr_2} . \tag{8.90}$$

Fluid properties of mercury and potassium are taken from Vargaftik and Touloukian (1975) and Dwyer (1976). The bubble departure diameter is the characteristic length used for Nu and Ra_2 but Eq. (8.90), like other buoyancy-driven correlations based on the 1/3-power law, is independent of any characteristic length. Fig. 8.4 shows the correlation of the experimental data with Eq. (8.90). New data is needed for a more reliable correlation.

CHAPTER 9

RADIATION

In this chapter, following a brief review of some radiative concepts, we develop radiation affected microscales of convection and, in terms of these scales, construct models for radiation affected convective heat transfer.

9.1 RADIATIVE STRESS

Spectrally averaged definitions of the radiative internal energy, heat flux and stress in terms of intensity I are

$$u^R = \frac{1}{c} \int_\Omega I \, d\Omega = \frac{1}{c} J \,, \tag{9.1}$$

$$q_i^R = \int_\Omega I \ell_i \, d\Omega \,, \tag{9.2}$$

$$\tau_{ij}^R = \frac{1}{c} \int_\Omega I \ell_i \ell_j \, d\Omega = \frac{1}{c} \Pi_{ij} \,, \tag{9.3}$$

where the J-scalar and the Π_{ij}-tensor are introduced for notational convenience, c is the velocity of light, ℓ_i is the unit vector in the direction of intensity, and Ω is the solid angle. In terms of these definitions, the first three specular moments of the transfer equation are

$$\frac{\partial q_i^R}{\partial x_i} = \kappa_P (B - J) \,, \tag{9.4}$$

176

$$\frac{\partial \Pi_{ij}}{\partial x_j} = -\kappa_R q_i^R ,\qquad(9.5)$$

$$\Pi_{ij} = \frac{1}{3} B \delta_{ij} + \sum_{n=1}^{\infty} \frac{1}{\kappa_M^{2n}} \left(M_{ijpq} \cdots \frac{\partial}{\partial x_p} \frac{\partial}{\partial x_q} \right) B ,\qquad(9.6)$$

with

$$M_{ijpq} \cdots = \frac{1}{4\pi} \int_{\Omega} (\ell_i \ell_j \ell_p \ell_q \cdots) \, d\Omega .\qquad(9.7)$$

Here $B = 4E_b$, $E_b = \sigma T^4$ being the Stefan-Boltzmann law for the black body emissive power, κ_P and κ_R are the Planck and Roseland means of the absorption coefficient, respectively, and $\kappa_M = (\kappa_P \kappa_R)^{1/2}$ is the geometric mean of these coefficients. The incorporation of κ_P and κ_R into the foregoing equations is discussed by Traugott (1966), Cogley et al. (1968), and their use in a variety of problems by Arpacı and coworkers (see Arpacı, 1991 and the references cited therein). Clearly, Eq. (9.4) denotes the thermal balance, Eq. (9.5) the momentum balance associated with radiation, and Eq. (9.6) gives the Π_{ij}-tensor in terms of a series based on the specular moments defined by Eq. (9.7). Note that the radiative heat flux given by Eq. (9.5), rearranged as

$$q_i^R = -\frac{1}{\kappa_R} \frac{\partial \Pi_{ij}}{\partial x_j} ,\qquad(9.8)$$

can be interpreted as a generalized diffusion process for any optical thickness. A procedure for the evaluation of Eq. (9.7) in terms of the Wallis integrals is described in Unno and Spiegel (1966). After lengthy manipulations, this procedure leads to

$$\Pi_{ij} = \sum_{n=0}^{\infty} \frac{\nabla^{2n-2}(2n \partial_i \partial_j + \nabla^2 \delta_{ij}) B}{\kappa_M^{2n}(2n+1)(2n+3)} ,\qquad(9.9)$$

where $\partial_i = \partial/\partial x_i$ and $\partial_j = \partial/\partial x_j$ are used for notational convenience. The same result may be found also in earlier works (see, for example, Milne, 1930). The formal similarity of Eq. (9.9) to the Hookean constitution for elastic solids should be noted (see Arpacı, 1991).

An alternate form for this stress may be given in terms of the isotropic radiative pressure. First, invoking the assumption of isotropy, Eqs. (9.1) and (9.3) are related as

$$\tau_{ij}^R = \frac{1}{3} u^R \delta_{ij} ,\qquad(9.10)$$

which implies

$$\Pi_{kk} = J , \tag{9.11}$$

where

$$\frac{1}{3c}\Pi_{kk} = -p \tag{9.12}$$

is the (isotropic) pressure of radiation. Then, the trace of Π_{ij}, in view of $\ell_k \ell_k = 1$, gives

$$\Pi_{kk} = \sum_{n=0}^{\infty} \left(\frac{\nabla^2}{\kappa_M^2}\right)^n \frac{B}{(2n+1)} . \tag{9.13}$$

Now, in a manner similar to the incorporation of the isotropic pressure into the development of viscous stress from elastic stress, (see, for example, Arpacı and Larsen, 1984), adding the identity

$$\frac{1}{3}J\delta_{ij} - \frac{1}{3}\Pi_{kk}\delta_{ij} = 0 \tag{9.14}$$

to Eq. (9.9), the Π_{ij}-tensor may be rearranged in terms of the radiation pressure,

$$\Pi_{ij} = \frac{1}{3}J\delta_{ij} + \sum_{n=0}^{\infty} \frac{2n\nabla^{2n-2}(\partial_i\partial_j - \frac{1}{3}\nabla\delta_{ij})B}{\kappa_M^{2n}(2n+1)(2n+3)} . \tag{9.15}$$

The formal similarity of Eq. (9.15) to the viscous (Stokesean) stress and the electromagnetic (Maxwell) stress should be noted (see Arpacı, 1991). This similarity is to be expected in view of the assumed isotropy for the elastic, viscous and electromagnetic continua (see, for example, Stratton, 1941 and Prager, 1961). The use of the first term of Eq. (9.15) in place of Eq. (9.9) is the well-known Eddington approximation which leads to a diffusive heat flux,

$$q_i^R = -\frac{1}{3\kappa_R}\frac{\partial J}{\partial x_i} , \tag{9.16}$$

for any optical thickness. The maximum deviation of this flux from the exact flux given by Eq. (9.8) is about 29% at $\tau = 1/\sqrt{3}$ (see Arpacı, 1991). Insertion of Eq. (9.16) into Eq. (9.4) gives a first order approximate balance of radiative energy,

$$(\nabla^2 - 3\kappa_M^2)J = -12\kappa_M^2 E_b . \tag{9.17}$$

The next section deals with dimensional arguments for radiation.

9.2   QUALITATIVE RADIATION

On dimensional grounds, Eq. (9.16) becomes

$$q^R \sim \frac{\Delta J}{3\kappa_R \delta} , \qquad (9.18)$$

δ being a diffusive length, ΔJ the change of J over this length. To relate J to temperature, consider the radiative constitution given by Eq. (9.17). By the help of Fourier transforms, for example,

$$\exp(ik_k x_j) , \qquad (9.19)$$

k_j being the wave number vector,

$$\nabla^2 = -k_0^2 , \quad k_0^2 = k_1^2 + k_2^2 + k_3^2 ,$$

or, in view of $k_0 \sim \delta^{-1}$,

$$\nabla^2 \sim -\delta^{-2} ,$$

and Eq. (9.17) yields

$$(\delta^{-2} + 3\kappa_M^2)J \sim 12\kappa_M^2 E_b . \qquad (9.20)$$

Then, in terms of optical thickness

$$\tau \sim \kappa_M \delta , \qquad (9.21)$$

$$J \sim \left(\frac{12\tau^2}{1 + 3\tau^2} \right) E_b \qquad (9.22)$$

which, together with Eq. (9.18), leads to the radiative heat flux,

$$q^R \sim 4\eta \left(\frac{\tau}{1 + 3\tau^2} \right) \Delta E_b , \qquad (9.23)$$

valid for any optical thickness but excludes any boundary effect. Here, $\eta = (\kappa_P/\kappa_R)^{1/2}$ denotes a measure for nongrayness. For $\eta = 1$, Fig. 9.1 shows the radiative heat flux versus emission and absorption which respectively are the measures for hotness and thickness of gas.

Equation (9.23) excludes any effect of scattering. To include this effect, consider the thick gas limit with isotropic scattering,

$$q_i^R = -\frac{4}{3\beta_R} \frac{\partial E_b}{\partial x_i} , \quad \tau \to \infty \qquad (9.24)$$

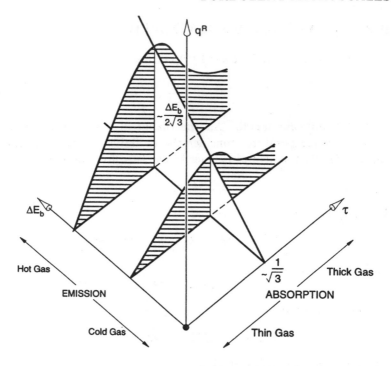

FIGURE 9.1: Radiative heat flux versus emission and absorption.

where $\beta_R = \kappa_R + \sigma_R$, β_R and σ_R respectively being the Rosseland mean of extinction and scattering coefficients. On dimensional grounds, and in terms of the albedo of isotropic scattering,

$$\omega = \sigma_R / \beta_R , \qquad (9.25)$$

Eq. (9.24) yields

$$q^R \sim \frac{4\eta(1 - \omega)\Delta E_b}{3\tau} , \quad \tau \to \infty . \qquad (9.26)$$

Also, noting that the thin gas limit is independent of any scattering, Eq. (9.23) may be rearranged to include for $\tau \to \infty$ the effect of isotropic scattering. Then,

$$q^R \sim 4\eta \left[\frac{\tau}{1 + 3\tau^2/(1 - \omega)} \right] \Delta E_b . \qquad (9.27)$$

Figure 9.2 shows the radiative heat flux versus absorption and scattering.

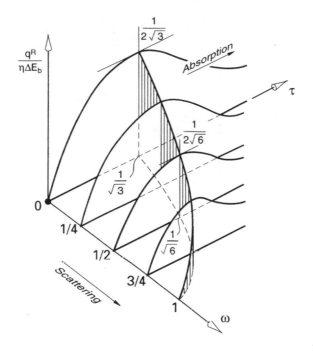

FIGURE 9.2: Radiative heat flux versus absorption and scattering.

To include any wall effect into Eq. (9.27), first consider the boundary affected thick gas and thin gas approximations with negligible scattering. For thick gas, from Arpacı and Larsen (1969) or Arpacı (1991),

$$q_y^R = -\frac{4}{3\kappa_R}\left(1 - \rho_w E_3 - \frac{3}{2}E_4\right)\frac{\partial E_b}{\partial y} \, ,\tag{9.28}$$

where ρ_w is the wall reflectivity, E_3 and E_4 are the exponential integrals of order three and four. Far from boundaries,

$$\lim_{y\to\infty}\left(\frac{E_3}{E_4}\right)\to 0 \, ,\tag{9.29}$$

and Eq. (9.28) is reduced to the Rosseland flux,

$$q_y^R\big|_\infty = -\frac{4}{3\kappa_R}\frac{\partial E_b}{\partial y} \, .\tag{9.30}$$

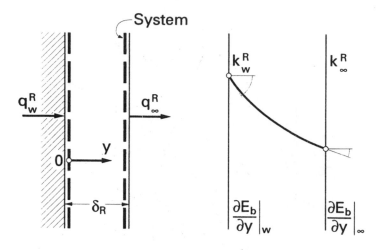

a) Radiative Balance   **b) Sink Effect**

FIGURE 9.3: Thick gas.

On boundaries,

$$E_3(0) = \frac{1}{2}, \quad E_4(0) = \frac{1}{3},$$
(9.31)

and Eq. (9.28) gives

$$q_y^R\Big|_\infty = -\frac{4}{3\kappa_R} \left(\frac{\epsilon_w}{2}\right) \frac{\partial E_b}{\partial y}\Big|_w,$$
(9.32)

where ϵ_w is the wall emissivity. The diffusive nature of thick gas is well-known. However, across a thin radiative boundary layer of thickness $\delta_R \sim \kappa_R^{-1}$, the radiative energy balance for the system shown in Fig. 9.3,

$$q_w^R = q_\infty^R,$$
(9.33)

in view of the fact that radiative conductivity

$$k_w^R < k_\infty^R,$$
(9.34)

leads to

$$\frac{\partial E_b}{\partial y}\Big|_w > \frac{\partial E_b}{\partial y}\Big|_\infty.$$
(9.35)

That is, the thick gas next to a wall acts as a radiative sink. On dimensional grounds,

$$q_w^R \sim \frac{4\eta}{3\tau} \left(\frac{\epsilon_w}{2}\right) \Delta E_b , \qquad (9.36)$$

where $\tau = \alpha_M y$ and y is the distance from wall.

For thin gas, from Lord and Arpacı (1970) or Arpacı (1991),

$$\frac{\partial q_y^R}{\partial y} = 4\kappa_P \left[(E_b - E_{b\infty}) - \frac{\epsilon_w}{2} (E_{bw} - E_{b\infty}) E_2 \right] , \qquad (9.37)$$

where E_2 is the exponential integral of order two. The behavior of thin gas as a radiative sink is well-known. On boundaries,

$$E_2(0) = 1 , \qquad (9.38)$$

and Eq. (9.37) becomes

$$\left.\frac{\partial q_y^R}{\partial y}\right|_w = 4\kappa_P \left(1 - \frac{\epsilon_w}{2}\right) (E_{bw} - E_{b\infty}) , \qquad (9.39)$$

or, on dimensional grounds, following an integration,

$$q_w^R \sim 4\eta\tau \left(1 - \frac{\epsilon_w}{2}\right) \Delta E_b + C , \qquad (9.40)$$

where $C = \epsilon_w \Delta E_b$ shows the effect of enclosure radiation.

Finally, in terms of Eqs. (9.23), (9.27), (9.36) and (9.40), the radiative flux including the effect of a hot wall, as well as that of emission, absorption and scattering, is

$$q_w^R \sim \left[\frac{4\eta \left(1 - \frac{\epsilon_w}{2}\right) \tau + \epsilon_w}{1 + 3 \left(\frac{2}{\epsilon_w} - 1\right) \tau^2/(1 - \omega)} \right] \Delta E_b . \qquad (9.41)$$

Also, for later convenience, we introduce a heat transfer number as a measure for radiation relative to conduction,

$$H = \frac{q_w^R}{q_w^K} \qquad (9.42)$$

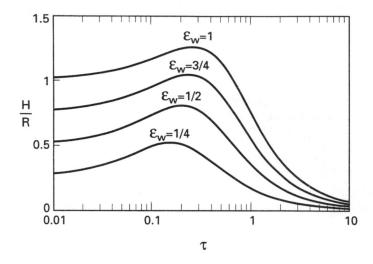

FIGURE 9.4: H/R versus τ.

which gives, in terms of Eq. (9.41),

$$H = \epsilon_w \left[\frac{1 + 2\eta \left(\frac{2}{\epsilon_w} - 1 \right) \tau}{1 + 3 \left(\frac{2}{\epsilon_w} - 1 \right) \tau^2/(1 - \omega)} \right] R \ , \qquad (9.43)$$

where

$$R \sim \frac{\text{Emission}}{\text{Conduction}} \sim \frac{\Delta E_b}{k \Delta T / \delta_\theta} \qquad (9.44)$$

is a Planck number, δ_θ the radiation affected thermal boundary layer. Fig. 9.4 shows H/R versus τ for $\eta = 1$, $\omega = 0$ and various values of ϵ_w. The foregoing review is utilized in the next section dealing with radiation affected thermal microscales.

9.3 FORCED FLOW

Under the influence of radiation, Eq. (4.89) is expanded to

$$\mathcal{P}_\theta \sim \left(\epsilon_\theta^K + \epsilon_\theta^R \right) \ , \qquad (9.45)$$

where the production retains its original form given by Eq. (4.88) but the dissipation becomes

$$\rho c_p \left(\epsilon_\theta^K + \epsilon_\theta^R \right) \sim \overline{\left(q_i^K + q_i^R \right) \left(\frac{\partial \theta}{\partial x_i} \right)} \tag{9.46}$$

which includes now radiative as well as conductive dissipation.

On dimensional grounds involving an integral scale ℓ and a radiation affected thermal scale λ_θ^R, Eq. (9.46) yields

$$\rho c u_\theta \frac{\theta}{\ell} \sim q_w^K \left(1 + \frac{q_w^R}{q_w^K} \right) \frac{\theta}{\lambda_\theta^R} , \tag{9.47}$$

or, in terms of Eq. (9.41), noting the *sink effect of radiation* and assuming $\delta_\theta = \lambda_\theta^R$,

$$u_\theta \frac{\theta^2}{\ell} \sim \alpha(1 - H) \frac{\theta^2}{(\lambda_\theta^R)^2} \tag{9.48}$$

Also, from Eq. (4.77),

$$u \sim \nu \frac{\ell}{\lambda^2} . \tag{9.49}$$

Now, for a small effect of radiation, neglecting departure from the similarity between velocity and temperature fields,

$$\frac{u}{\lambda} \sim \frac{u_\theta}{\lambda_\theta^R} , \tag{9.50}$$

which gives in terms of Eqs. (9.48) and (9.50)

$$\frac{\lambda}{\lambda_\theta^R} \sim \left(\frac{Pr}{1 - H} \right)^{1/3} \tag{9.51}$$

or,[1]

$$\frac{\lambda}{\lambda_\theta^R} \sim Pr^{1/3}(1 + H)^{1/3} \tag{9.52}$$

which, in the limit of the homogeneous flow, leads to

$$\frac{\lambda}{\lambda_\theta^R} \to \frac{\eta}{\eta_\theta^R} , \tag{9.53}$$

---

[1]Note $(1 - x)^{-1} \sim (1 + x)$ for a small x.

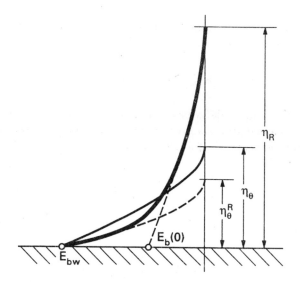

FIGURE 9.5: Various boundary layers.

or,

$$\frac{\eta}{\eta_\theta^R} \sim Pr^{1/3}(1+H)^{1/3} \,, \tag{9.54}$$

or,

$$\frac{\eta}{\eta_\theta^R} \sim Pr_R^{1/3} \,, \tag{9.55}$$

where Pr_R is a radiation-affected Prandtl number,

$$Pr_R = (1+H)Pr \,, \tag{9.56}$$

Pr being the usual Prandtl number. Fig. 9.5 is a sketch on various boundary layers, η_R corresponding to pure radiation.

In terms of the Kolmogorov scale,

$$\eta \sim \left(\frac{\nu^3}{\epsilon}\right)^{1/4} \,, \tag{4.53}$$

Eq. (9.54) leads to a thermal microscale,

$$\eta_\theta^R \sim \frac{\eta_\theta}{(1+H)^{1/3}} \,, \tag{9.57}$$

where η_θ is the mesomicroscale introduced, in Chapter 4, as an average of the Kolmogorov and Batchelor scales (recall Eq. 4.114).

Next, we proceed to radiation affected microscales of buoyancy-driven flows.

9.4   BUOYANCY-DRIVEN FLOW

For buoyancy-driven homogeneous flow of fluids with $Pr \geq 1$, consider the reduced balance of the mean kinetic energy of fluctuations (recall Eq. 4.131)

$$\mathcal{B} \sim \epsilon \qquad (9.58)$$

coupled with the radiation-affected balance of the mean of thermal fluctuations,

$$\mathcal{P}_\theta \sim \epsilon_\theta^K + \epsilon_\theta^R . \qquad (9.45)$$

On dimensional grounds, Eqs. (9.58) and (9.45) respectively lead to

$$\mathcal{B} \sim \nu \frac{u^2}{\lambda^2} \sim \epsilon \qquad (9.59)$$

and, in view of Eq. (9.48) for $H \ll 1$, to

$$u_\theta \sim \frac{\alpha \ell}{(1 + H)(\lambda_\theta^R)^2} . \qquad (9.60)$$

Extending the Squire postulate by assuming

$$\lambda \sim \lambda_\theta^R , \qquad (9.61)$$

recalling $u \sim u_\theta$ for buoyancy-driven flows, and eliminating velocity between Eqs. (9.59) and (9.60), lead to a radiation-affected thermal Taylor scale

$$\lambda_\theta^R \sim \ell^{1/3} \left[\frac{\nu \alpha}{(1 + H)^2 \mathcal{B}} \right]^{1/6} \qquad (9.62)$$

whose limit for

$$\left(\frac{\lambda_\theta^R}{\ell} \right) \to \left(\eta_\theta^B \right)^R , \qquad (9.63)$$

is a radiation-affected Batchelor scale,

$$\left(\eta_\theta^B\right)^R \sim \left[\frac{\nu\alpha^2}{(1+H)\mathcal{B}}\right]^{1/4} . \tag{9.64}$$

In terms of buoyant force

$$\mathcal{F} \sim \mathcal{B}/u_\theta \tag{9.65}$$

and local velocity

$$u_\theta \sim \alpha/(1+H)\left(\eta_\theta^B\right)^R , \tag{9.66}$$

Eq. (9.64) becomes

$$\left(\eta_\theta^B\right)^R \sim \left[\frac{\nu\alpha}{(1+H)\mathcal{F}}\right]^{1/3} , \tag{9.67}$$

or, relative to an integral scale ℓ,

$$\frac{\left(\eta_\theta^B\right)^R}{\ell} \sim Ra_R^{-1/3} , \tag{9.68}$$

where Ra_R is a radiation-affected Rayleigh number,

$$Ra_R = (1+H)Ra , \tag{9.69}$$

Ra being the usual Rayleigh number,

$$Ra = \frac{\mathcal{F}\ell^3}{\nu\alpha} . \tag{9.70}$$

Having developed radiation-affected microscales for forced and buoyancy-driven flows, we proceed to the final section of the chapter in which we construct microscale based heat transfer models for these flows.

9.5   HEAT TRANSFER

The Nusselt number, defined in terms of an integral scale and a thermal Kolmogorov scale,

$$Nu \sim \frac{\ell}{\eta_\theta} , \tag{9.71}$$

after assuming $\eta_\theta \sim \eta_\theta^R$, rearranged in terms of the Kolmogorov scale η,

$$Nu \sim \left(\frac{\ell}{\eta}\right)\frac{\eta}{\eta_\theta^R} \, , \tag{9.72}$$

applies to forced convection. Then, in terms of Eqs. (4.64) and (9.55),

$$Nu \sim Re_\ell^{3/4} Pr_R^{1/3} \, , \tag{9.73}$$

Pr_R being the radiation-affected Prandtl number,

$$Pr_R = (1 + H)Pr \, , \tag{9.56}$$

Pr the usual Prandtl number and H the heat transfer number defined by Eq. (9.43).

The Nusselt number, after assuming $\eta_\theta \sim (\eta_\theta^B)^R$,

$$Nu \sim \frac{\ell}{(\eta_\theta^B)^R} \tag{9.74}$$

applies to natural convection. Then, in terms of Eq. (9.68),

$$Nu \sim Ra_R^{1/3} \, , \tag{9.75}$$

Ra_R being the radiation-affected Rayleigh number,

$$Ra_R = (1 + H)Ra \, , \tag{9.76}$$

Ra the usual Rayleigh number.

In the absence of radiation, the Nusselt number given by Eqs. (9.73) and (9.75) agrees well with the experimental literature and with earlier empirical correlations. At present, experimental data on radiation-affected turbulent convection are not available.

CHAPTER 10

EQUILIBRIUM SPECTRA

Having accomplished the major objectives of the monograph, that is, microscales of complex flows and heat and mass transfer models in terms of these scales, we proceed to equilibrium spectra of these flows. Forced flows leading to wave number dependence of $\kappa^{-5/3}$ for kinetic spectra and to that of $\kappa^{-5/3}$ and κ^{-1} for thermal spectra are well-known (Fig. 10.1). See, for example, Tennekes and Lumley (1972). Here, we consider only the buoyancy-driven and pulsating flow spectra which are early candidates for experimental interest.

10.1 BUOYANCY-DRIVEN FLOW

10.1.1 Energy (Kinetic) Spectra

The isotropic limit of Eq. (4.133), after invoking the Squire postulate, gives

$$\mathcal{B} \sim \frac{v^3}{\eta_\theta} + \nu \frac{v^2}{\eta_\theta^2} . \tag{10.1}$$

First, consider low Prandtl number fluids. Rearrange Eq. (10.1) relative to inertial production as

$$\mathcal{B} \sim \frac{v^3}{\eta_\theta} \left(1 + \frac{\nu}{v \eta_\theta} \right) , \tag{10.2}$$

190

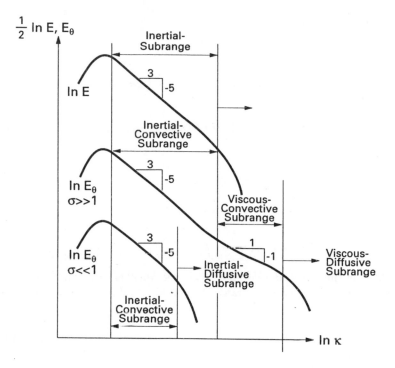

FIGURE 10.1: Kinetic and thermal spectra of fluids with large and small Prandtl numbers.

or, in terms of the isotropic limit of Eq. (4.134), as[1]

$$B \sim \frac{v^3}{\eta_\theta}(1+\sigma) \tag{10.3}$$

which gives

$$v \sim \left(\frac{B}{1+\sigma}\right)^{1/3} \eta_\theta^{1/3} \tag{10.4}$$

and the kinetic energy

$$E \sim v^2 \sim \left(\frac{B}{1+\sigma}\right)^{2/3} \eta_\theta^{2/3}. \tag{10.5}$$

---

[1]In this chapter, we customarily use σ for the Prandtl number.

By definition,

$$E \sim \int_0^\infty E(\kappa) \, d\kappa \tag{10.6}$$

$E(\kappa)$ being the energy spectra and κ the wave number. Over a narrow (equilibrium) range,

$$E \sim \kappa E(\kappa) . \tag{10.7}$$

After letting

$$\eta_\theta \sim \kappa^{-1} \tag{10.8}$$

in Eq. (10.5), combine the result with Eq. (10.7) to get

$$E(\kappa, \sigma) \sim \left(\frac{B}{1+\sigma} \right)^{2/3} \kappa^{-5/3}, \quad \sigma \ll 1 , \tag{10.9}$$

which is the *inertial subrange* of kinetic spectra for liquid metals. As $\sigma \to 0$ (recall Eqs. 4.144-4.148), Eq. (10.9) is reduced to the Kolmogorov spectra,

$$E(\kappa, 0) \sim \epsilon^{2/3} \kappa^{-5/3}, \quad \sigma \to 0 . \tag{10.10}$$

Eq. (10.9) relative to Eq. (10.10) gives

$$\frac{E(\kappa, \sigma)}{E(\kappa, 0)} = \left(\frac{B/\epsilon}{1+\sigma} \right)^{2/3}, \quad \sigma \ll 1 \tag{10.11}$$

where $1 \le B/\epsilon \le 2$.

Next, consider higher Prandtl number fluids and rearrange Eq. (10.1) relative to viscous dissipation as

$$B \sim \nu \frac{v^2}{\eta_\theta^2} \left(1 + \frac{\nu \eta_\theta}{\nu} \right) , \tag{10.12}$$

or, in terms of the isotropic limit of Eq. (4.134), as

$$B \sim \nu \frac{v^2}{\eta_\theta^2} \left(1 + \frac{1}{\sigma} \right) \tag{10.13}$$

which gives

$$v \sim \left(\frac{B}{\nu} \right)^{1/2} \left(1 + \frac{1}{\sigma} \right)^{-1/2} \eta_\theta \tag{10.14}$$

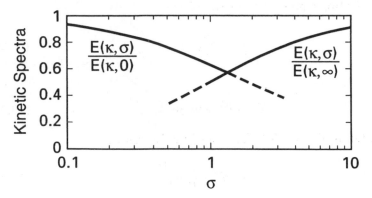

FIGURE 10.2: Kinetic spectra versus σ for $B/\epsilon = 1$.

and the kinetic energy

$$E \sim v^2 \sim \left(\frac{B}{\nu}\right)\left(1 + \frac{1}{\sigma}\right)^{-1} \eta_\theta^2 . \tag{10.15}$$

In view of Eq. (10.8), elimination of E between Eqs. (10.7) and (10.15) yields

$$E(\kappa, \sigma) \sim \left(\frac{B}{\nu}\right)\left(1 + \frac{1}{\sigma}\right)^{-1} \kappa^{-3} , \quad \sigma > 1 \tag{10.16}$$

which is the *viscous subrange* of kinetic spectra for liquids including water and viscous oils. As $\sigma \to \infty$ (recall Eqs. 4.140, 4.142), Eq. (10.16) is reduced to

$$E(\kappa, \infty) \sim \left(\frac{\epsilon}{\nu}\right) \kappa^{-3} , \quad \sigma \to \infty . \tag{10.17}$$

Eq. (10.16) relative to Eq. (10.17) gives

$$\frac{E(\kappa, \sigma)}{E(\kappa, \infty)} = \left(\frac{B}{\epsilon}\right)\left(\frac{\sigma}{1 + \sigma}\right) , \quad \sigma > 1 \tag{10.18}$$

Fig. 10.2 shows Eqs. (10.11) and (10.18) versus σ for $B/\epsilon = 1$.

10.1.2   Thermal (Entropy) Spectra[2]

Rearrange the temperature variance,

$$E_\theta \sim \vartheta^2 , \tag{10.19}$$

---

[2]The temperature variance of spectra studies.

by the isotropic production of Eq. (4.134),

$$v\frac{\vartheta^2}{\eta_\theta} \sim \alpha\frac{\vartheta^2}{\eta_\theta^2} \sim \epsilon_\theta \, , \tag{10.20}$$

to get

$$E_\theta \sim \vartheta^2 \sim \eta_\theta\epsilon_\theta v^{-1} \, , \tag{10.21}$$

or, in terms of Eq. (10.4) for small Prandtl numbers,

$$E_\theta \sim \vartheta^2 \sim \epsilon_\theta \left(\frac{1+\sigma}{B}\right)^{1/3} \eta_\theta^{2/3} \, . \tag{10.22}$$

By definition,

$$E_\theta \sim \int_0^\infty E_\theta(\kappa)\, d\kappa \, , \tag{10.23}$$

and, over a narrow range,

$$E_\theta \sim \kappa E_\theta(\kappa) \, . \tag{10.24}$$

In view of Eq. (10.8), elimination of E_θ between Eqs. (10.22) and (10.24) yields

$$E_\theta(\kappa,\sigma) \sim \epsilon_\theta \left(\frac{1+\sigma}{B}\right)^{1/3} \kappa^{-5/3} \, , \quad \sigma \ll 1 \tag{10.25}$$

which is the *inertial-convective* subrange of thermal spectra for liquid metals. As $\sigma \to 0$, Eq. (10.25) is reduced to

$$E_\theta(\kappa,0) \sim \epsilon_\theta\epsilon^{1-/3}\kappa^{-5/3} \, , \quad \sigma \to 0 \, . \tag{10.26}$$

Eq. (10.25) relative to Eq. (10.26) gives

$$\frac{E_\theta(\kappa,\sigma)}{E_\theta(\kappa,0)} = \left(\frac{\epsilon}{B}\right)^{1/3}(1+\sigma)^{1/3} \, , \quad \sigma \ll 1 \, . \tag{10.27}$$

Next, combine Eqs. (10.14) and (10.21), eliminate E_θ between the result and Eq. (10.24), and note Eq. (10.8) to get

$$E_\theta(\kappa,\sigma) \sim \epsilon_\theta \left(\frac{\nu}{B}\right)^{1/2}\left(1+\frac{1}{\sigma}\right)^{1/2}\kappa^{-1} \, , \quad \sigma > 1 \tag{10.28}$$

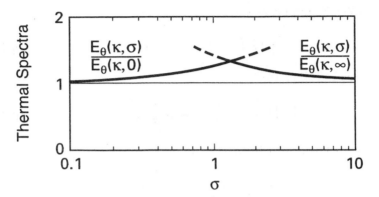

FIGURE 10.3: Thermal spectra versus σ for $B/\epsilon = 1$.

which is the *viscous-convective* subrange of thermal spectra for liquids including water and viscous oils. As $\sigma \to \infty$,

$$E_\theta(\kappa, \infty) \sim \epsilon_\theta \left(\frac{\nu}{\epsilon}\right)^{1/2} \kappa^{-1} , \quad \sigma \to \infty . \tag{10.29}$$

Eq. (10.28) relative to Eq. (10.29) gives

$$\frac{E_\theta(\kappa, \sigma)}{E_\theta(\kappa, \infty)} = \left(\frac{\epsilon}{B}\right)^{1/2} \left(1 + \frac{1}{\sigma}\right)^{1/2} , \quad \sigma > 1 . \tag{10.30}$$

Fig. 10.3 shows Eqs. (10.27) and (10.30) versus σ for $B/\epsilon = 1$. Finally, rearrange Eq. (10.19) by the isotropic dissipation of Eq. (4.134) to get

$$E_\theta \sim \vartheta^2 \sim \left(\frac{\epsilon_\theta}{\alpha}\right) \eta_\theta^2 , \tag{10.31}$$

or, in terms of Eqs. (10.8) and (10.24),

$$E_\theta(\kappa) \sim \left(\frac{\epsilon_\theta}{\alpha}\right) \kappa^{-3} \tag{10.32}$$

which is the *conductive subrange* of thermal spectra.

10.2   PULSATING FLOW

Rearrange the isotropic limit of Eq. (7.13),

$$\omega v^2 + \frac{v^3}{\eta} \sim \nu \frac{v^2}{\eta^2} \sim \epsilon , \tag{10.33}$$

relative to spatial production,

$$\frac{v^3}{\eta}\left(1 + \frac{\omega\eta}{v}\right) \sim \nu\frac{v^2}{\eta^2} \sim \epsilon \,, \tag{10.34}$$

or, in view of Eq. (7.18),

$$\frac{v^3}{\eta}\left[1 + \omega\left(\frac{\nu}{\epsilon}\right)^{1/2}\right] \sim \nu\frac{v^2}{\eta^2} \sim \epsilon \,. \tag{10.35}$$

Then, an inertial estimate for velocity is

$$v \sim \left[\frac{\epsilon\eta}{1 + \omega\left(\frac{\nu}{\epsilon}\right)^{1/2}}\right]^{1/3} \tag{10.36}$$

and the kinetic energy in terms of this velocity is

$$E \sim v^2 \sim \left[\frac{\epsilon\eta}{1 + \omega\left(\frac{\nu}{\epsilon}\right)^{1/2}}\right]^{2/3} . \tag{10.37}$$

Elimination of E between Eqs. (10.7) and (10.37) gives, in view of

$$\eta \sim \kappa^{-1} \,, \tag{10.38}$$

$$E(\kappa, \omega) \sim \frac{\epsilon^{2/3}\kappa^{-5/3}}{\left[1 + \omega\left(\frac{\nu}{\epsilon}\right)^{1/2}\right]^{2/3}} \,, \qquad \omega\left(\frac{\nu}{\epsilon}\right)^{1/2} < 1 \,. \tag{10.39}$$

As $\omega \to 0$, Eq. (10.39) is reduced to the Kolmogorov spectra given by Eq. (10.10). Eq. (10.39) relative to Eq. (10.10) gives

$$\frac{E(\kappa, \omega)}{E(\kappa, 0)} = \frac{1}{\left[1 + \omega\left(\frac{\nu}{\epsilon}\right)^{1/2}\right]^{2/3}} \,, \qquad \omega\left(\frac{\nu}{\epsilon}\right)^{1/2} < 1 \,. \tag{10.40}$$

Next, rearrange Eq. (10.33) relative to temporal production,

$$\omega v^2\left(1 + \frac{v}{\omega\eta}\right) \sim \nu\frac{v^2}{\eta^2} \sim \epsilon \,. \tag{10.41}$$

Then, another inertial estimate for velocity is, in view of Eq. (7.18),

$$v \sim \left[\frac{\epsilon/\omega}{1 + \left(\frac{\epsilon}{\nu}\right)^{1/2}/\omega} \right]^{1/2} \tag{10.42}$$

and kinetic energy

$$E \sim v^2 \sim \left[\frac{\epsilon/\omega}{1 + \left(\frac{\epsilon}{\nu}\right)^{1/2}/\omega} \right] \tag{10.43}$$

which gives, in terms of Eqs. (10.7) and (10.38),

$$E(\kappa,\omega) \sim \frac{(\nu\epsilon)^{1/2}\kappa^{-1}}{\left[1 + \omega\left(\frac{\nu}{\epsilon}\right)^{1/2}\right]}, \qquad \omega\left(\frac{\nu}{\epsilon}\right)^{1/2} > 1. \tag{10.44}$$

As $\omega \to \infty$, Eq. (10.44) is reduced to the Stokes spectra,

$$E(\kappa) \sim \left(\frac{\epsilon}{\omega}\right)\kappa^{-1}. \tag{10.45}$$

Eq. (10.44) relative to Eq. (10.45) gives

$$\frac{E(\kappa,\omega)}{E(\kappa,\infty)} = \frac{\omega\left(\frac{\nu}{\epsilon}\right)^{1/2}}{1 + \omega\left(\frac{\nu}{\epsilon}\right)^{1/2}}, \qquad \omega\left(\frac{\nu}{\epsilon}\right)^{1/2} > 1. \tag{10.46}$$

Fig. 10.4 shows Eqs. (10.40) and (10.46) versus $\omega(\nu/\epsilon)^{1/2}$. Finally, a viscous estimate of velocity, obtained from Eq. (10.33), is

$$v \sim \eta\left(\frac{\epsilon}{\nu}\right)^{1/2} \tag{10.47}$$

and kinetic energy

$$E \sim v^2 \sim \left(\frac{\epsilon}{\nu}\right)\eta^2, \tag{10.48}$$

combined with Eqs. (10.7) and (10.34), yields for the *viscous subrange*,

$$E(\kappa) \sim \left(\frac{\epsilon}{\nu}\right)\kappa^{-3} \tag{10.49}$$

which is independent of ω, as expected (recall Eq.10.17).

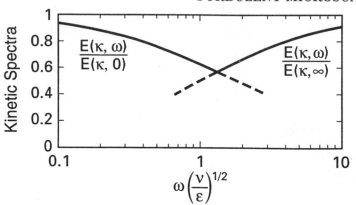

FIGURE 10.4: Kinetic spectra versus $\omega(\nu/\epsilon)^{1/2}$.

PROBLEMS

10-1) Develop the equilibrium spectra of the following turbulent flows:

  i) Rotating

 ii) Thermo-capillary driven

iii) Reacting

iv) Pulsating

  v) Two-Phase

vi) Radiation affected

EPILOGUE

Two roads diverged in a wood and I —
I took the one less traveled by, ...

FROST

Since the discovery of microscales of turbulence by Kolmogorov and their extensions by Oboukhov, Corrsin and Batchelor, research on *universal* aspects of turbulence is dormant. Entire contemporary research is devoted to *structure* of turbulent flows. The present monograph is alone among this research in its attempt to originate a general approach for microscales of complex turbulent flows and to demonstrate their relevance to heat and mass transfer correlations.

REFERENCES

Alhaddad, A. A. and Coulman, G. A., 1982, "Experimental and theoretical study of heat transfer in pulse-combustion heaters," *Proceedings Vol.I: Symposium on Pulse Combustion Applications*, GRI-82/0009.2, Atlanta, GA.

Alpert, R. L., 1977, "Pressure modeling of fires controlled by radiation," *Sixteenth Symposium (International) on Combustion*, p. 1489. The Combustion Institute, Pittsburgh.

Arpacı, V. S., 1966, *Conduction Heat Transfer*, Addison-Wesley, Reading, Massachusetts.

Arpacı, V. S., 1986, "Microscales of turbulence and heat transfer correlations," *Int. J. Heat Mass Transfer*, Vol. 29, p. 1071.

Arpacı, V. S., 1987, "Radiative entropy production–lost heat into entropy", *Int. J. Heat Mass Transfer*, Vol. 30, p. 2115.

Arpacı, V. S., 1990a, "Microscales of turbulence and heat transfer correlations," *Annual Review of Heat Transfer*, Vol. 3, p. 195, (Edited by C. L. Tien), Hemisphere Publishing Corporation.

Arpacı, V. S., 1990b, "Foundations of entropy production", *Advances in Thermodynamics*, Vol. 3, p. 369 (Eds. S. Sieniutycz and P. Salomon), Taylor and Francis, New York.

Arpacı, V. S., 1991, "radiative entropy production-heat loss into entropy," *Advances in Heat Transfer*, Vol. 21, p. 239, (Edited by Hartnett, J. P. and Irvine, T. F.). Academic Press, New York.

Arpacı, V. S., 1994a (Keynote Lecture), "Microscales of turbulence, mass transfer correlations," *International Symposium on Turbulence, Heat and Mass Transfer*, Lisbon, Portugal.

Arpacı, V. S., 1994b (Keynote Lecture), "Microscales of turbulence, heat transfer correlations", *Tenth International Heat Transfer Conference*, Brighton, England.

Arpacı, V. S., 1995a, "Buoyant turbulent flow driven by internal energy generation," *Int. J. Heat Mass Transfer*, Vol. 38, p. 2761.

Arpacı, V. S., 1995b, "Microscales of turbulent combustion," *Prog. Energy Combust. Sci.*, Vol. 21, p. 153.

Arpacı, V. S. and Bayazıtoglu, Y., 1973, "Thermal stability of radiating fluids: Asymmetric slot problem", *Phys. Fluids*, Vol. 16, p. 589.

Arpacı, V. S. and Dec, J. E. 1987, "A theory for buoyancy driven turbulent flows," *24th National Heat Transfer Conf.*, Paper No. 87-HT-55, Pittsburgh, Pennsylvania (August 1987).

Arpacı, V. S., Dec, J. E. and Keller, J. O., 1993, "Heat transfer in pulse combustor tailpipes," *Combust. Sci. and Tech.*, Vol. 94, p. 131.

Arpacı, V. S. and Gözüm, D., 1973, "Thermal stability of radiation fuids: The Benard problem", *Phys. Fluids*, Vol. 16, p. 581.

Arpacı, V. S., Guessous, L., Dec, J. E., and Keller, J. O., 1996, "Heat transfer enhanced by pulsation," presented at the Bergles Symposium, Georgia Institute of Technology, Atlanta.

Arpacı, V. S. and Kao, S.-H., 1996, "Microscales of turbulent two-phase film," presented at the 1996 National Heat Transfer Conference, Houston.

Arpacı, V. S. and Kao, S.-H., 1997, "Foundations of buoyancy driven heat transfer correlations," under preparation.

Arpacı, V. S. and Larsen, P. S., 1969, "A thick gas model near boundaries", *AIAA J.*, Vol. 7, p. 602.

Arpacı, V. S. and Larsen, P. S., 1984, *Convection Heat Transfer*, Prentice Hall, Englewood Cliffs, New Jersey.

Arpacı, V. S. and Larsen, P. S., 1988, "Microscales of turbulent two-phase flows," in ASME, HTD-96, Vol. 2, p. 423.

Arpacı, V. S. and Li, C. Y., 1995, "Turbulent forced diffusion flames," *Comb. Flame*, Vol. 102, p. 170.

Arpacı, V. S. and Selamet, A., 1988, "Entropy production in flames", *Combust. flame*, Vol. 73, p. 251.

Arpacı, V. S. and Selamet, A., 1991, "Buoyancy driven turbulent diffusion flames," *Combust. Flame*, Vol. 86, p. 203.

Arpacı, V. S., Selamet, A., Chai, A. T., and Duh, J. C., 1990, "Multiscale dimensional foundations of surface tension- and/or buoyancy-driven two-phase flows," *Proceedings of Seventh European Symposium on Materials and Fluid Sciences in Microgravity*, Oxford, UK, p. 485.

Arpacı, V. S. and Tabaczynski, R. J., 1982, "Radiation-affected flame propagation", *Combust. Flame*, Vol. 46, p. 315

Arpacı, V. S. and Tabaczynski, R. J., 1984, "Radiation-affected laminar flame quenching", *Combust. Flame*, Vol. 57, p. 169.

Arpacı, V. S. and Troy, S., 1990, "On the attenuating thin gas", *J. Thermophysics Heat Transfer*, Vol. 4, p. 407.

Avedisian, C. T., Ioffredo, C., and O'Connor, M. J., 1984, "Film boiling of discrete droplets of mixtures of coal and water on a horizontal brass surfaces," *Chem . Eng. Sci.*, Vol. 39, p. 319.

Batchelor, G. K., 1959, "Small-scale variation of convected quantities like temperature in a turbulent fluid," *J. Fluid Mech.*, Vol. 5, p. 113.

Block, M. J., 1956, "Surface tension as the cause of Bénard cells and surface deformation in a liquid film," *Nature*, Vol. 178, p. 650.

Bromley, L. A., 1950, "Heat transfer in stable film boiling," *Chem. Eng. Prog.*, Vol. 46, p. 221.

Bui, T. D. and Dhir, V. K., 1985, "Film boiling heat transfer on an isothermal vertical surface," *ASME J. Heat Transfer*, Vol. 107, p. 764.

Burgess, D. S., Strasser, A., and Grumer, J., 1961, *Fire Res. Abst. Rev.*, Vol. 3, p. 177.

Busemann, A., 1933, *Der Wärme-und Stoffaustausch. Springer, Berlin.*

Busse, F. H., 1981, "Transition to turbulence in rayleigh-benard convection," in *Hydrodynamic Instabilities and the Transition to Turbulence* (H. L. Swinney and J. P. Gollub, Eds.), Vol. 45, p. 97, Springer-Verlag.

Chandrasekhar, S., 1961, *Hydrodynamic and Hydromagnetic Stability*. Oxford University Press.

Cheung, F. B., 1980, "Heat source-driven thermal convection at arbitrary Prandtl number," *J. Fluid Mech.*, Vol. 97, p. 743.

Chu, T. Y., 1993, "A correlational approach to turbulent saturated film boiling," *ASME J. Heat Transfer*, Vol. 115, p. 993.

Chu, T. Y. and Goldstein, R. J., 1973, "Turbulent convection in a horizontal layer of water," *J. Fluid Mech.*, Vol. 60, p. 141.

Churchill, S. W. and Chu, H. H. S., 1975, "Correlating equations for laminar and turbulent free convection from a vertical plate," *Int. J. Heat Mass Transfer*, Vol. 18, p. 1323.

Cogley, A. C., Vincenti, W. G., and Gilles, S. E., 1968, "Differential approximation for radiative transfer in a nongrey gas near equilibrium", *AIAA J.*, Vol. 6, p. 551.

Corlett, R. C., 1968, "Gas fires with pool-like boundary conditions," *Combust. Flame*, Vol. 12, p. 19.

Corlett, R. C., 1970, "Gas fires with pool-like boundary conditions: Further results and interpretation," *Combust. Flame*, Vol. 14, p. 351.

Corrsin, S., 1951, "On the spectrum of isotropic temperature fluctuations in isotropic turbulence," *J. Appl. Phys.*, Vol. 22, p. 469.

Corrsin, S., 1962, "Turbulent dissipation fluctuations," *Phys. Fluids*, Vol. 5, p. 1301.

de Ris, J. and Orloff, L., 1972, "A dimensionless correlation of pool burning data," *Combust. Flame*, Vol. 18, p. 381.

Dec, J. E. and Keller, J. O., 1989, "Pulse combustor tail-pipe heat-transfer dependence on frequency, amplitude, and mean flow rate," *Combustion and Flame*, Vol. 77, p. 359.

Dec, J. E., Keller, J. O. and Arpacı, V. S., 1992, "Heat transfer enhancement in the oscillating turbulent flow of a pulse combustor tailpipe," *Int. J. Heat Mass Transfer*, Vol. 35, p. 2311.

Drazin, P. G. and Reid, W. H., 1981, *Hydrodynamic Stability*. Cambridge University Press.

Dwyer, O. E., 1976, *Boiling Liquid-Metal Heat Transfer*, p. 421. American Nuclear Society, Hinsdale, Illinois.

Elder, J. W., 1969, "The temporal development of a model of high Rayleigh number convection," *J. Fluid Mech.*, Vol. 35, p. 417.

Emmons, H. W., 1956, "The film combustion of liquid fuel," *Z. Angew. Math. Mech.*, Vol. 36, p. 60.

Fiedler, H. and Wille, R, 1971, "Warmetransport bie frier Konvektion in einer horizontalen Flussigkeitsschicht mit Volumenheizung, Teil 1: Integraler Warmetransport," *Rep. Dtsch Forschungs-Varsuchsanstalt Luft-Raumfahrt, Inst. Turbulenzfenschung*, Berlin.

Fitzjarrald, D. E., 1976, "An experimental study of turbulent convection in air," *J. Fluid Mech.*, Vol. 73, p. 693.

Frederkin, T. H. K. and Clark, J. A., 1962, "Natural convection film boiling on a sphere," *Advances in Cryogenic Engineering*, Vol. 8, p. 501.

Frederkin, T. H. K. and Daniels, D. J., 1966, "The relation between bubble diameter and frequency of removal from a sphere during film boiling," *J. Heat Transfer*, Vol. 88, p. 87.

Frisch, U., 1995, *Turbulence. The legacy of A. N. Kolmogorov*. Cambridge University Press.

Fujii, T. and Imura, H., 1972, "Natural convection heat transfer from a plate with arbitrary inclination," *Int. J. Heat Mass Transfer*, Vol. 15, p. 755.

Galitseyskiy, B. M. and Ryzhov, Y. A., 1977, "Heat transfer in turbulent gas flows in the case of high-frequency pressure fluctuations," *Heat Transfer-Soviet Research*, Vol. 8, No. 4, July-Aug., p. 178.

Garon, A. M. and Goldstein, R. J., 1973, "Velocity and heat transfer measurements in thermal convection," *Phys. Fluids*, Vol. 16, p. 1818.

Gill, J., 1967, "Interferometric measurement of temperature gradient reversal in a layer of convecting air," *J. Fluid Mech.*, Vol. 30, p. 371.

Globe, S. and Dropkin, D., 1959, "Natural convection heat transfer in liquids confined by two horizontal plates and heated from below," *Trans. Am. Soc. Mech. Engrs., J. Heat Transfer C.*, Vol. 81, p. 24.

Goldstein, R. J. and Chu, T. Y., 1969, "Thermal convection in a horizontal layer of air," *Prog. Heat and Mass Transfer*, Vol. 2, p. 55-75.

Goldstein, R. J. and Tokuda, S., 1980, "Heat transfer by thermal convection at high rayleigh numbers," *Int. J. Heat Mass Transfer*, Vol. 23, p. 738.

Goody, R. M., 1964, *Atmospheric Radiation*, Oxford.

Gorkow, L. P., 1958, "Stationary convection in a plane liquid layer near the critical heat transfer point," *Soviet Physics, JETP*, Vol. 6, p. 311.

Haas, J. C., Arpacı, V. S., and Springer, G. S., 1971, "Mass and heat transfer in a diatomic gas", *J. Plasma Physics*, Vol. 6, p. 547.

Hanby, V. I., 1969, "Convective heat transfer in a gas- fired pulsating combustor," *ASME J. of Engr. for Power*, Vol. 91, p. 48.

Herring, J. R., 1963, "Investigation of problems in thermal convection," *J. Atmos. Sci.*, Vol. 20, p. 325.

Hussain, A. K. M. F and Reynolds, W. C., 1970, "The mechanics of an organized wave in turbulent shear flow," *J. Fluid Mech.*, Vol. 41, p. 241.

Hussain, A. K. M. F and Reynolds, W. C., 1972, "The mechanics of an organized wave in turbulent shear flow. Part 2. Experimental results," *J. Fluid Mech.*, Vol. 54, p. 241.

Kanury, A. M., 1975, "Modeling of pool fires with a variety of polymers," *Fifteenth Symposium (International) on Combustion*, p. 193. The Combustion Institute, Pittsburgh.

Kanury, A. M., 1977, *Introduction to Combustion Phenomena.* Gordon and Breach, New York.

Kim, J. S., de Ris, J., and Kroesser, F. W., 1971, "Laminar free convective burning of fuel surfaces," *Thirteenth Symposium (International) on Combustion*, p. 949. The Combustion Institute, Pittsburgh.

Koh, J. C. Y., Sparrow, E. M., and Hartnett, J. P., 1961, "Two phase boundary layer in laminar film condensation," *Int. J. Heat Mass Transfer*, Vol. 2, p. 69.

Kolmogorov, A. N., 1941, "Local structure of turbulence in incompressible viscous fluid for very large Reynolds numbers," *C. R. Acad. Sci. USSR*, Vol. 30, p. 301.

Krishnamurti, R., 1970a, "On the transition to turbulent convection. Part 1. The transition from two- to three-dimensional flow," *J. Fluid Mech.*, Vol. 42, p. 295.

Krishnamurti, R., 1970b, "On the transition to turbulent convection. Part 2. The transition to time-dependent flow," *J. Fluid Mech.*, Vol. 42, p.309.

Krishnamurti, R., 1973, "Some further studies on the transition to turbulent convection," *J. Fluid Mech.*, Vol. 60, p.285.

Kulacki, F. A. and Emara, A. A., 1977, "Steady and transient thermal convection in a fluid layer with uniform volumetric energy sources," *J. Fluid Mech.*, Vol. 83, p. 375.

Kulacki, F. A. and Nagle, M. E., 1975, "Natural convection in a horizontal fluid layer with volumetric energy sources," *J. Heat Transfer*, Vol. 91, p. 204.

Kuriyama, T., 1987, *Saturated film-boiling heat transfer on vertical surface*, Ph.D. Thesis, University of Tokyo, Tokyo, Japan.

Landau, L. D., 1944, "On the problem of turbulence," *C. R. Doklady Acad. Sci. URSS*, Vol. 44, p. 311.

Liao, N. S., Wang, C. C., and Hong, J. T., 1985, "An investigation of heat transfer in pulsating turbulent pipe flow," ASME 23rd National Heat Transfer Conference, Denver, Colorado. *Fundamentals of Forced and Mixed Convection*, editors Kulacki, F. A. and Boyd, R. D., HTD-Vol. 42.

Liaw, S. and Dhir, V. K., 1986, "Effect of surface wettability on transition boiling heat transfer from a vertical surface," *Proceedings of Eighth International Heat Transfer Conference*, Vol. 4, p. 2031. University of California, Los Angeles, C.A.

Lockwood, R. W. and Corlett, R. C., 1987, "Radiative and convective feedback heat flux in small turbulent pool fires with variable pressure and ambient oxygen," *Heat and Mass transfer in Fires*, HTD-73, p. 421.

Long, R. R., 1976, "Relation between Nusselt number and Rayleigh number in turbulent thermal convection," *J. Fluid Mech.*, Vol. 73, p. 445.

Lord, H. and Arpacı, V. S., 1970, "Effect of nongray thermal radiation on laminar forced convection over a heated horizontal plate", *Int. J. Heat Mass Transfer*, Vol. 13, p. 1737.

Lyon, R. N., 1952, *Liquid Metals Handbook*, 3rd Ed., Washington, D. C.: Atomic Energy Commission and Department of the Navy.

Malkus, W., 1954, "The heat transport and spectrum of thermal turbulence," *Proc. Roy. Soc. (London) A*, Vol. 225, p. 196.

Malkus, W. and Veronis, G., 1958, "Finite amplitude cellular convection," *J. Fluid Mech.*, Vol. 4, p. 225.

Marxman, G. A., 1967, "Boundary-layer combustion in propulsion," *Eleventh Symposium (International) on Combustion*, p. 269. The Combustion Institute, Pittsburgh.

Marxman, G. A. and Gilbert, M., 1963, "Turbulent boundary layer combustion in the hybrid rocket," *Ninth Symposium (International) on Combustion*, p. 371. The Combustion Institute, Pittsburgh.

Milne, E. A., 1930, "Thermodynamics of stars", *in Handbuch der Astrophysik*, Vol. 3, Chap. 2, p. 65.

Nakagawa, Y., 1960, "Heat transport by convection," *Physics of Fluids*, Vol. 3, p. 82.

Nishio, S. and Chandratilleke, G. R., 1989, "Steady-state pool boiling heat transfer to saturated liquid helium at atmospheric pressure," *JSME Int. J. II*, Vol. 32, p. 639.

Nishio, S., Chandratilleke, G. R., and Ozu, T., 1991, "Natural-convection film-boiling heat transfer (saturated film boiling with long vapor film)," *JSME Int. J. II*, Vol. 34, p. 202.

Nusselt, W., 1916, "Die Oberflachenkondensation des Wasser Dampfes," *Zeitschr. V D I*, Vol. 60, p. 541.

Oboukhov, A. M., 1949, "Structure of the temperature field in turbulent flows," *Izv. Nauk. SSSR, Geogr. i. Geofiz.*, Vol. 13, p. 58.

Paul, P. J., Mukunda, H. S., and Jain, V. K., 1982, *Nineteenth Symposium (International) on Combustion*, The Combustion Institute, Pittsburgh, pp. 717.

Pearson, J. R. A., 1958, "On convection cells induced by surface tension," *J. Fluid Mech.*, Vol. 4, p. 489.

Phillips, W. F. and Arpacı, V. S., 1972, "Diatomic gas-thermal radiation interaction: A model equation for the internal fluid", *J. Plasma Physics*, Vol. 7, p. 235.

Phillips, W. F. and Arpacı, V. S., 1975, "Monatomic plasma-thermal radiation interaction: A weakly ionized kinetic model", *J. Plasma Physics*, Vol. 13, p. 523.

Phillips, W. F., Arpacı, V. S., and, Larsen, P. S., 1970, "Diatomic gas-thermal radiation interaction between parallel plates", *J. Plasma Physics*, Vol. 4, p. 429.

Prager, W., 1961, *Introduction to Mechanics of Continua*, p. 87.

Reynolds, W. C. and Hussain, A. K. M. F, 1972, "The mechanics of an organized wave in turbulent shear flow. Part 3. Theoretical models and comparison with experiments," *J. Fluid Mech.*, Vol. 54, p. 263.

Ribaud, G., 1941, "Nouvelle expression du coefficient de convection de la chaleur en regime decoulment turbuelnt," *J. Phys. Radium*, Vol. 2, p. 12.

Schmidt, E. and Beckmann, W., 1930, *Tech. Mech. Thermodynam. Berl.*, Vol. 1, p. 341.

Silveston, P. L. 1958, "Warmedurchgang in waagerechten flussigkeit-schichten," *Forsch-Geb. Ing-Nes.*, Vol. 29, p. 59.

Somerscales, E. F. C. and Gazda, I. W., 1969, "Thermal convection in high Prandtl number liquids at high Rayleigh number," *Int. J. Heat Mass Transfer*, Vol. 12, p. 1491.

Spalding, D. B., 1954, "Mass transfer in laminar flow," *Proc. Roy. Soc.*, Vol. A221, p. 78.

Sparrow, E. M. and Gregg, J. L., 1959, "A Boundary layer treatment of laminar film condensation," *J. Heat Transfer*, Vol. 81, p. 13.

Squire, H. B., 1938, " Free convection from a heated vertical plate," *Modern developments in fluid mechanics* (ed. Goldstein), Oxford, Vol. 2, p. 638.

Stratton, J. A., 1941, *Electromagnetic Theory*, McGraw-Hill, New York.

Stuart, J. T., 1958, "On the nonlinear mechanics of hydrodynamic stability," *J. Fluid Mech.*, Vol. 4, p. 1.

Suryanarayana, N. V. and H. Merte, H., 1972, "Film boiling on vertical surfaces," *ASME J. Heat Transfer* Vol. 94, p. 377.

Taylor, G. I., 1935, "Statistical theory of turbulence," *Proc. Roy. Soc. A*, Vol. 151, p. 421.

Tennekes, H., 1968, "Simple model for the small scale structure of turbulence," *Phys. Fluids*, Vol. 11, p. 669.

Tennekes, H. and Lumley, O. L., 1972, *A First Course in Turbulence*, MIT Press.

Thomas, D. B. and Townsend, A. A., 1957, "Turbulent convection over a heated horizontal surface," *J. Fluid Mech.*, Vol. 2, p. 473.

Threlfall, D. C., 1975, "Free Convection in Low-Temperature Gaseous Helium," *J. Fluid Mech.*, Vol. 67, p. 17.

Tritton, D. J. and Zarraga, M. N., 1967, "Convection in horizontal layers with internal heat generation," *J. Fluid Mech.*, Vol. 30, p. 21.

Traugott, S. C., 1966, "Radiative heat-flux potential for a nongrey gas", *AIAA J.*, Vol. 4, p. 541.

Unno, W. and Spiegel, E. A., 1966, "The Eddington approximation in the radiative heat equation", *Publications of the Astronomical Society of Japan*, Vol. 18, p. 85.

Vargaftik, N. B. and Touloukian, Y. S., 1975, *Handbook of Physical Properties of Liquids and Gases: Pure Substances and Mixtures* (2nd Edn). Hemisphere Publishing Corporation.

Veronis, G., 1959, "Cellular convection with finite amplitude in a rotating fluid," *J. Fluid Mech.*, Vol. 5, p. 401.

Wachters, L. H. J., Bonne, H., and van Nouhuis, H. J., 1966, "The heat transfer from a hot horizontal plate to sessile water drops in the spheroidal state," *Chem. Eng. Sci.*, Vol. 21, p. 923.

INDEX